P9-CBL-492

THE GHOSTS *of* EVOLUTION

Also by Connie Barlow

From Gaia to Selfish Genes: Selected Writings in the Life Sciences (Editor)

Evolution Extended: Biological Debates on the Meaning of Life (Editor)

Green Space, Green Time: The Way of Science

CONNIE BARLOW

THE GHOSTS *of* EVOLUTION

Nonsensical Fruit, Missing Partners, and Other Ecological Anachronisms

A Member of the Perseus Books Group

Published by Basic Books,
A Member of the Perseus Books Group

A Cataloging-in-Publication record for this book is available from the Library of Congress.
ISBN: 0–465–00551–9

Designed by Jeffrey P. Williams
Set in 11 point Berkeley

FIRST EDITION

01 02 03 / 10 9 8 7 6 5 4 3 2 1

To all who remember

CONTENTS

FOREWORD

After five decades of probing the mysteries of large animals that vanished at the end of the Pleistocene (the end of the last ice age), I've emerged with some unusual ideas. For example, it's impossible for me to ignore the fact that America today is incomplete. When we lost mammoths, mastodons, and gomphotheres, not to mention native horses, ground sloths of all sizes, and dozens of other genera of wonderful large animals, we lost our American Serengeti. This happened only 13,000 years ago, long after the extinction of the dinosaurs. It was more than an injury; it was an ecological and evolutionary amputation. I dream of a memorial, a cenotaph, or a special mass to commemorate the disaster.

Few if any plants departed with the megafauna, but plants too have suffered the loss, as we shall see. A shared interest in such matters led to a phone call from science writer Connie Barlow, who decided to pursue this secret of Pleistocene ecology when my attempts at some simple answers proved insufficient. Here is the result. Focusing not on the extinct but on the partners they left behind, Connie offers us an intriguing Pleistocene field guide and brings us face to face with a mysterious quid pro quo, the ancient agreement between large animals and plants.

The plant side of the story can be seen at a glance in any good supermarket. The exemplar of what I mean is a ripe avocado, its nutritious pulp enveloping a large, inedible or mildly poisonous, ready-to-germinate seed. Other examples include mangoes, papaya, tamarind, peaches, and watermelon. All cater to fruit lovers, animals ranging in size from monkeys up to elephants. The payback to the plant for feeding the animal is dispersal of its seed. No free-market economist should be surprised at this transaction. Yet the animals that coevolved with the avocado are lost in the mists of time.

Years ago, ecologist Dan Janzen was the first to show that many of the larger fallen fruits of the tropics are going nowhere. In Costa Rica Dan ran feeding trials with some of the largest native animals found in tropical forests, such as tapir. They were too small to disperse the seeds of the large fruits. Daring to experiment with domestic livestock (what self-respecting ecologist would stoop to study the food habits of a *cow* in dry tropical forest?), Dan also discovered that although cattle, horses, and mules are large enough to ingest and disperse many of the larger fruits, they can, at best, suggest solutions, not solve the problem. They are alien newcomers that did not coevolve with the tropical trees of the New World.

In the absence of domestic cattle and horses, it is likely that since the extinction of the great beasts, fallen fruit piled up under calabash, guanacaste, honey locust, and many other trees. Small beetle larvae, many other seed predators, rot, and mildew wasted the crop.

Could this be true? In 1979 Dan Janzen wrote to me about a wild idea he had. I loved it.

He sent a draft and asked me to coauthor a manuscript, "Neotropical Anachronisms: The Fruits the Gomphotheres Ate." I added a few more names to Dan's list of extinct prehistoric animals, and we proudly submitted our paper to the flagship periodical for all who think they have discovered great new stuff: *Science*.

It was quickly rejected.

I was ready to quit and try elsewhere, but Dan battled an outraged editor who implied that our paper sounded too much like a Hollywood script to grace the prestigious pages of *Science*. I don't know how Dan managed to get the editors to reconsider. Finally they did, and that's where the story of the gomphotheres and the anachronistic fruits first appeared.

No matter where they are published or what their perceived merit, most scientific articles have a short shelf life. I figured I'd seen the last of "Neotropical Anachronisms." So you can imagine how thrilled I am that twenty years later Dan's and my paper has generated a blizzard of citations. Connie Barlow has now brought this literature together in a wide-ranging story that captures, better than anyone has before, the profound influence Dan's idea has had. It illustrates the value of adding a much-needed and often-neglected dimension to ecology: the dimension of time.

Here in her book Connie shows why alternative explanations for the mystery of the fallen fruits are insufficient if not flawed. She brings in Gary Nabhan's idea of how humans could have helped the plants disperse their fruits in the millennia after the native megafauna had vanished. She finds that in city parks, in botanical gardens, at field stations, and in warm temperate woodlands, as well as in the dry and wet tropical forests of the world, Dan's wild idea is gaining traction. She adds some astute ideas of her own. The natural history here is state-of-the-art.

Toward the end Connie tells of an event that I think was the first of its kind, reflecting the spirit of the times. In the summer of 1999, with the support of staff of the Mammoth Site at Hot Springs, South Dakota, Connie helped me organize an unusual requiem, a memorial for the extinct mammoths, mastodons, and gomphotheres. The audience of mourners heard from Georges Cuvier, the first paleontologist to show that mammoths and mastodons were extinct; from Thomas Jefferson, who hoped that they weren't; from geologist Larry Agenbroad, who claimed that frozen tissue from mammoths in Siberia just might yield enough DNA to clone a living woolly mammoth; and from me. I claimed that in the absence of our native American elephants ecologists do not begin to know the nuts and bolts of vegetation dynamics in this hemisphere.

The service ended with a moving soliloquy from a Ms. Honey Locust of New York City. Ms. Locust sported a big green hat bedecked with long, spiraling pods. Honey locust trees may be streetwise, we learned, but their pods miss the megafauna.

Come to think of it, that *Science* editor was right. As Connie Barlow shows us, beyond its merit as an ecological model, the mystery of fallen fruit makes a splendid script.

Paul S. Martin
Emeritus Professor of Geosciences
The Desert Laboratory
The University of Arizona, Tucson

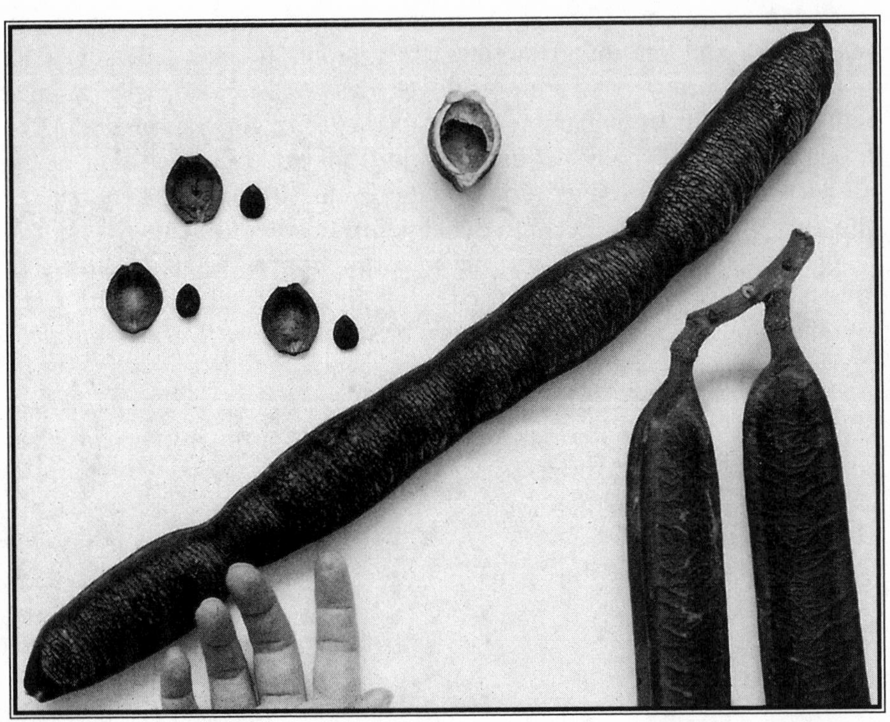

The Costa Rican fruit that launched a research program. The woody pods of *Cassia grandis* contain seeds embedded in a molasses-like pulp. Three seeds are shown, along with the thin walls that divide the pod into single-seed chambers. Cross section of a pod, cleaned of its seeds and pulp, is at top center. Author's hand for scale.

1

Ghost Stories

We live on a continent of ghosts, their prehistoric presence hinted at by sweet-tasting pods of mesquite, honey locust, and monkey ear.

—Paul Martin

"I've got a screwy idea," wrote Dan Janzen in an October 1977 letter to a colleague. "Let's write a paper together. I sit in my deciduous forest in the Pacific lowlands of Costa Rica looking at a large number of fairly large fleshy fruits with hard seeds. . . . An enormous number of these fruits end up simply rotting on the forest floor. When cattle are introduced, they busily go around and gobble up quantities of these fruits, but even then they sometimes don't get all of them. I think you can see where I'm going."

Where Janzen was going was toward a profound shift in ecological thinking.

Dan Janzen is a tropical ecologist based at the University of Pennsylvania. Costa Rica has been his site for fieldwork since the mid-1960s. He is the sort of guy who attracts the Crawfoord Prize (biology's highest honor) while still in his forties. It took me most of a morning to read his publications list. At the time Janzen was seeking a collaborator for his "screwy idea," he was thirty-eight years old and solidly established in his field. Plant-animal interactions in the tropics were the focus of his work. In 1977, however, he was beginning to suspect that he—along with every other ecologist working with large tropical fruits of the New World—had been wrong in one very big way. They all had failed to see that some fruits are adapted primarily for animals that have been extinct for thirteen thousand years.

Janzen was beginning to suspect that forests and savannas where livestock grazed were in some ways closer approximations to landscapes of the prehuman past than were nature preserves where cattle and horses were banned. The clue was rotting fruit.

What are fruits for, after all? Fruits—big or small, and produced by all flowering plants—contain the seeds of the next generation. Every seed is made of an embryo provisioned with a sack lunch. The sprouting plant will draw upon the stored energy and nutrients in the seed until it can capture sunlight in its own leaves and acquire minerals through its roots. Yet there is more to a fruit than a seed. Almost all fruits are equipped as well with a way for the kids to leave home. A maple seed is given wings, a dandelion tufts. A fleshy fruit like an apple, orange, grape, or tomato is given a lure. The lure is the pulp, and its mission is to induce a mobile animal of the correct specifications to serve as the vehicle for the seed or seeds within.

Janzen's letter to his hoped-for collaborator continues, "How hard would it be to make a list of the big mammals that would have eaten the same size and kind of fruits that are eaten by peccaries, tapirs, deer, and cows, and that would have occurred in a Pacific coastal lowland deciduous forest within the last 100,000 years?"

Peccaries, tapirs, and deer are the largest herbivores native to Costa Rica today. These are the animals that can eat the biggest and the most fruits, but they do not serve as vehicles for many of the seeds hidden within the lures. Some seeds are too big; a tapir or deer will eat around the seed or spit it out before swallowing. Other seeds may be crushed in the molar mill. The piglike peccaries are especially formidable seed predators. As well, too many fruits may fall from a tree all at once. Even though the coterie of local dispersal agents might be inclined to consume the fallen fruits properly, molds and microbes make a meal of them first. Still other fruits fail to find a proper vehicle because they remain out of reach when ripe.

Janzen's mention of cows may seem odd, because cattle are not native to the Western Hemisphere. They were introduced from Europe. Nevertheless, they now serve the reproductive interests of some tropical plants better than the natives can. Janzen's view of cattle had shifted away from the standard ecological portrayal of livestock as alien invaders. Rather, he had come to think of cattle as proxies for the great beasts that had inhabited not just Costa Rica but the entire Western Hemisphere for all but the final thirteen thousand of the last fifty million years.

This vast chunk of evolutionary time is the Cenozoic era, commonly known as the Age of Mammals. Many tropical fruits of the Americas evolved their size, color, odor, seed casing, and other adaptive features during millions of years of association with big mammals. We all know that the Age of Dinosaurs (correction: the Age of *Nonavian* Dinosaurs) ended sixty-five million years ago, when one or more big meteors struck the planet. It is a matter of taste whether one wishes to regard the Age of Mammals as over now too. There should be no disagreement, however, that the Age of *Great* Mammals has indeed ended everywhere except southern Africa and patches of tropical Asia.

The Age of Great Mammals ended long before chainsaws and internal combustion engines evolved. In Europe and nontropical regions of Asia, it petered out in steps between fifty and fifteen thousand years ago, when the straight-tusked elephants, woolly mammoths, rhinos, and other great beasts of the Pleistocene epoch vanished. Throughout that vast continental mass, a quarter of all genera of animals regarded as megafauna—those weighing more than a hundred pounds, or forty-five kilograms—were lost to extinction. Europe lost all six species of herbivores weighing more than a thousand kilograms. In Australia the Age of Great Mammals ended sometime between forty and thirty thousand years ago, when giant kangaroos, enormous wombats, and rhinolike marsupials (as well as the most formidable crocodiles, lizards, and snakes) were purged from the landscape. This extinction catastrophe stripped Australia of all but one of its sixteen genera of megafauna. In the Western Hemisphere, the Age of Great Mammals came to an abrupt end thirteen thousand years ago, when the mastodons and mammoths, the ground sloths and glyptodonts, the native horses and large camels, and a beaver and an armadillo both as big as a bear all disappeared forever.[1] North America lost 68 percent of its generic richness of Pleistocene megafauna (32 of 47 genera), and South America lost 80 percent (47 of 59 genera).[2]

Outlying islands were hit even harder, though several thousand years later, following advances in sailing technologies. Not until seven thousand years ago, for example, did Cuba lose its half dozen species of sloth, including a ground dweller as big as a black bear.[3] Just four thousand years ago, while the Egyptians were building pyramids, the last mammoths on Earth expired on an island off Siberia. Madagascar lost all of its biggest lemurs within the past two thousand years, along with both native species of hippopotamus, a strange carnivore, and the giant elephant birds unique to that island. Throughout the Cenozoic, New Zealand had no mammals other than bats, but less than a millennium ago colonizing peoples handily exterminated all dozen species of flightless moa.

Most of us live in impoverished places. Each year throngs of visitors ogle the wildlife in America's national parks, unaware that for tens of millions of years the land has never been so bereft of big creatures. The bison and elk and moose of Yellowstone are a remnant of what came before. We can still experience the great beasts that were lost, however, if we learn to read the plants.

The Age of Great Mammals may be over, but the plants have not yet caught on. Those that depended upon mammals to swallow big fruits, as well as those that deployed armaments to deter soft snouts from stripping foliage, are still doing what they have always done. Fruit rotting on the ground is the most obvious sign. Later, Janzen would notice other examples of adaptations mismatched in time. But in his initial letter, tropical fruits commanded his attention:

"The point is that if we were to take an East African coastal forest community and dump into it the fruits that fall into a Costa Rican deciduous forest community, it is my certain prediction that all of these fruits would be immediately eaten. . . . Now if we were to remove all the animals from an East African deciduous forest and leave only two species of pigs, one deer, and one tapir, how long would it take before the trees lost their mammal-dispersed fruits or otherwise adjusted themselves to the shortage of seed dispersal agents?"

Janzen continued, "I don't even know where to begin to try to work out a list of the sorts of animals that would have been available in these forests up until, say, ten or a hundred thousand years ago. That's what I am hoping you could do, perhaps right out of your head."

The *you* was Paul Martin. Martin is a Pleistocene ecologist, now emeritus at the University of Arizona in Tucson, and one of the leading experts on the Ice Age ecology of North America. Other scientists may know more about mammoths or ground sloths or glacial climates or vegetational shifts. Martin's specialty is Pleistocene generalist. His work on the ecological interactions between the megafauna of the Western Hemisphere and the first human colonists brought him notoriety a decade before Dan Janzen sought his assistance. Janzen wanted to draw upon Martin's knowledge of the Pleistocene bestiary and to solicit his opinion about likely megafaunal interactions—not with humans but with fruits.

Janzen's letter continued: "Obviously I haven't thought the thing all the way through, but I'm curious as to whether you would be interested in trying to put something together with me? For my part, I'll be happy to put together a list of the plants and what their fruits are like and documentation on their acceptability and edibility by deer, peccaries, and a tapir (I have a corralled tapir whom I have been feeding things to). I'll be happy to write a first draft of the manuscript as well. What I'd hope for from you would be some kind of a list of the animals that could have been involved in the evolution of these fruits, their supposed dietary habits, perhaps a statement of what extant animals they resemble most closely, and some high quality skepticism about the harebrained ideas I am sure will come to light as this thing actually gets organized on paper."

The letter concludes, "I think the whole thing that started me thinking about this was frustration over trying to figure out who the dispersal agents are for *Cassia grandis* seeds. The fruit is about a half meter long, hard as a rock, cylindrical, and an inch and a half in diameter, and contains large seeds (about 2 cm long by 1.5 cm wide by .5 cm thick). These seeds are embedded in a sweet molasses-like pulp, which is probably highly desirable to any animal that can break the fruit. The fruits hang on the tree long enough so that bruchid beetles and moths kill all the seeds. The inference is that under the circumstances of the evolution of this plant, some big mammal came along and picked the fruits out of the tree before the insects got all the seeds. A giant ground sloth comes immediately to mind as a possibility. You can see why I thought of you. Onward into battle. Sincerely yours, Dan."

Paul Martin responded promptly: "What a fun letter and idea. I'll invoke the ghosts of some hungry extinct herbivores and you will see if they eat up the fallen fruit. You write the paper, I propitiate the extinct spirits, and off it goes to some unsuspecting editor. Am I reading you OK? Despite your disclaimer, you have thought the thing through very well."

Right at the outset Martin spoke of the Pleistocene megafauna as ghosts. Ghosts are what Martin sees when he is interacting with plants, and other scientists have since referred to the phenomenon of missing partners in the same way.[4] Janzen, however, has steadfastly refrained from using that word. He prefers to focus on the living rather than the dead. By the time he wrote the first draft of their joint paper, Janzen would have an equally compelling way to characterize the plant side of the partnership: *anachronisms*.

Paul Martin's initial response to Dan Janzen is worth quoting at length because it is a fine introduction to the Pleistocene menagerie that will appear throughout this book. Like Janzen's letter, Martin's is delightful for its playfulness and unalloyed enthusiasm.

"You want a scenario for 100,000 B.C.? That should be the last interglacial. I won't apologize for the bum fossil record and poor dating in lowland tropics. Without consulting the literature, I come up with the following hungry mouths:

- Giant ground sloths, 18 feet tall, heavier than an African elephant . . . and bipedal at least some of the time. Genus *Eremotherium*. Teeth are like pruning shears.
- Some smaller ground sloths, either a *Mylodon*, a *Scelidotherium*, or both. Their teeth would be less well adapted for chopping than *Eremotherium*.

- Some kind of mastodont forest elephants, either *Cuvieronius* or *Stegomastodon*, or both. If the giant sloth didn't, these guys did slurp up the fallen fruit with their trunks and beat the bruchid beetles to the huge *Cassia* pods. The *Cassia* seeds get a fine, big turd of elephant manure to germinate in, just like *Acacia tortilis* in Africa.
- A rhino-sized monstrosity, *Toxodon*, the last of his order. Doris Stone, or somebody, once showed me *Toxodon* bones from the museum in San José. His mouth is full of big buck teeth incisors, like a huge mule. Great for bark shredding?
- One or more species of *Glyptodont*, their heads never far from the ground and their teeth too small for grinding. The largest were big enough to miscegenate with a Honda Civic.
- Some extinct peccaries, cf. *Platygonus*, maybe twice the weight of the white-lipped and even more rangy. *Catagonus* in Paraguay, their nearest living relative, isn't quite extinct yet.
- Possibly some tropical horse.
- Probably giant tortoises (genus *Geochelone*).

"My hunch is that the mastodonts got 70 to 80 percent of your fruit, and that the extinct megafauna had a carrying capacity in your area comparable to the Lake Manyara Reserve, which I think is over 100 animal units per section. [Lake Manyara is in Tanzania; an animal unit is a measurement developed by the livestock industry to indicate herbivory equivalent to one cow plus calf—about a thousand pounds, or 450 kilograms.]

"While you start the first draft," Paul Martin concluded his letter, "I'll dust off some books on who really lived in Costa Rica or nearby. At least you've solved the *Cassia grandis* problem. I'm assuming those pods are within easy elephant reach, say, 10–12 feet, and of course with *Eremotherium* you can go higher. As ever, Paul S. Martin."

FOUR YEARS WOULD PASS BEFORE Janzen and Martin's joint effort would appear in the prestigious journal *Science*.[5] During that time Martin was aggregating fossil information as to which beasts had in fact lived in Costa Rica toward the end of the Pleistocene. He also ventured into the countryside of Sonora, Mexico, where he had long been conducting ecological fieldwork. He wanted to look at the plants with new eyes, now that ghosts were a possibility. It was there that he began to develop an idea that would prove useful not only for ecologists working with Pleistocene ghosts but also for studying African elephants. He called this concept "the megafaunal dispersal syndrome."

Meanwhile, Janzen conducted field experiments in Costa Rica that would provide fruit-specific and other empirical grounding. His studies of tough-husked guanacaste and jicaro fruits and the beneficial or predatory passage of their seeds through the digestive systems of horses, cattle, tapir, and mice were reported in several journals commencing in 1981. This fieldwork and experimentation provided crucial support for the ground-breaking theoretical paper.

"Neotropical Anachronisms: The Fruits the Gomphotheres Ate" was published in *Science* in January 1982. Neotropical anachronisms are anachronisms of the New World tropics. The term applies to plants that some time in the past thirty or forty million years evolved fruits intended to attract very large mammals. The big beasts are gone, but the fruits remain. Year after year in the American tropics (and temperate climes too), trees and vines produce fruits that make little sense today. Some fruits simply rot on the ground beneath the parent plant. Others are raided by seed predators or plundered by pulp thieves. Whether rotted, raided, or plundered, viable seeds are rarely dispersed.

The plants not only remember the great mammals of the Pleistocene and before; they expect gomphotheres, ground sloths, toxodons, and their ilk to show up any day now. Thirteen thousand years is not enough time for plants to notice and genetically respond to the loss.

Twenty years ago Dan Janzen and Paul Martin launched what has become a new research program in the biological sciences. Their work has also contributed to our awed enjoyment of the living world. Nature enthusiasts as well as scientists who encounter the anachronism concept will have their experience of the plant—and animal—realm transformed.

Haunting the Wild Avocado

Grocery stores are excellent places to encounter ghosts. They lurk in the fruit section, feasting on anachronisms. Paul Martin, who has been honing his occult skills for a quarter century, thinks he's spotted ghosts among the apples and pears. I'm a neophyte, however, so I head straight for the tropical fruits.

Papaya is haunted by spectacular ghosts. Most impressive are the gomphotheres and ground sloths. The forty species of genus *Carica* are native to South and Central America. Only four are deemed edible by humans. The kind sold in grocery stores, *Carica papaya,* probably originated in Mexico. All species of *Carica,* including those shunned by humans, are haunted by ghosts with gapes large enough to take in the soft fruit whole, rather than bite by bite.

The avocado bin attracts ghost glyptodonts and toxodons as well as gomphotheres and ground sloths. Because almost all fifty species of genus

Persea are native to the tropics and subtropics of the Americas, one can surmise that the avocado genus honed its form in the Western Hemisphere. Not all of these species evolved with megafauna in mind. The kind of *Persea* that thrives along the Gulf Coast of the United States bears a fruit not much bigger than a blueberry.

Like papaya, the avocado found in grocery stores (smooth and rough-skinned varieties of *Persea americana*) has been haunted for thirteen thousand years. Many living frugivores, omnivores, and even carnivores are attracted to the oily pulp, but only an animal with a massive gullet will swallow the huge seed along with the flesh. The cultivated varieties of *Persea americana* have far thicker pulps surrounding the seed than does the ancestral stock, but the seed itself is virtually unchanged in girth.[6] From a functional and evolutionary perspective, avocado *intends* its fruits to be swallowed whole. That's how the species disperses its seed. The oily flesh is simply the lure. A parent tree could wish for no more desirable fate for its offspring than to have its seeds plopped into the world within steaming heaps of dung.

Whether growing in commercial orchards of southern California or forest fragments of the neotropics, domestic and wild avocado trees still expect giant mammals to stop by for the harvest. Wave upon wave of Cenozoic megafauna faithfully harvested avocado fruits, season upon season, for tens of millions of years. The identities of the dispersers shifted every few million years, but from an avocado's perspective, a big mouth is a big mouth and a friendly gut is a friendly gut. The passage of a trifling thirteen thousand years is too soon to exhaust the patience of genus *Persea*. The genes that shape fruits ideal for megafauna retain a powerful memory of an extraordinary mutualistic relationship. Embellished by our own scientific understanding, that memory would look something like this:

The scene is a tropical forest in Central America fifteen thousand years ago, and a giant has just arrived. Perhaps attracted by the scent of ripe pulp, a three-ton mother and her bear-size toddler approach a tree that shed its fruit crop a few days before. The visitors are ground sloths, whose closest living relatives are South American tree sloths, anteaters, and armadillos. *Eremotherium* looks like nothing alive today. Think of a bear crossed with a prairie dog or marmot and endowed with the bulk of an elephant. The adult sloth begins to sniff the carpet of fruits for the ripest specimens. Her agile offspring climbs a nearby tree for safety and also because, at this age, climbing is not only possible but irresistible. In a few years, the young sloth's tree-climbing days will be over. By then, an enormous bulk and powerful clawed forelimbs will suffice to ward off all but the most determined predators.

The mother finds a fruit that smells acceptable and tests it for softness between frontally toothless jaws. The whole fruit is then mashed between tongue and palate. The slippery seed slides easily down the animal's gullet, along with the nutritious pulp. Laxatives in the pulp ensure that the seed will complete its dark journey before digestive juices do it harm.

Other seeds follow. Before she is satiated, the sloth and her young depart. The adult sloth will balance the oily meal with leafy browse, thus keeping microbes happy in the vast fermentation vat of her gut. Tomorrow the pair will return to the same tree, dispersing seeds along the way. Or perhaps *Eremotherium* will choose a papaya tree instead. To feed on papaya, the great sloth will sit up on her haunches, using her sturdy tail for a third point of balance. She will choose the ripest pendulous fruit—all of which are borne on the trunk of the small tree. Her reach may exceed four or even five meters.

The sloth's limbs still show signs of arboreal ancestry. In shuffling from plant to plant, *Eremotherium* walks on the sides of her paws. The awkward gait may owe to phylogenetic inertia—an inability to evolve away from an established form. Perhaps, too, it owes to the survival advantage of inturned paws. An enhanced ability to climb when young should more than offset an inability to run later on. Or perhaps the anatomical quirk is necessary for the sloth to walk at all. *Eremotherium*'s front feet bear exceptionally long claws, as do the front feet of a relative that will survive the end-Pleistocene extinction: South America's giant anteater. The anteater walks on its knuckles, claws behind and curled skyward.

Meanwhile, in another part of the forest, one that is especially rich in avocado trees, a small herd of gomphotheres (genus *Cuvieronius*) approaches on an ancient trail. The herd has traveled tens of kilometers in the past three days, munching greenery along the way. The matriarch remembers the route. She remembers this avocado-rich valley and others throughout a vast region, as well as good places and times to find papaya, cherimoya, sapote, *Cassia grandis,* and many other treats. She learned these sites while following the lead of her mother, the former matriarch.

The gomphotheres arrive at the avocado grove when the fruits are near their prime. A half dozen elephantine trunks probe the carpet of green-brown fruits. This is the first avocado experience for the youngest member of the clan. The pulp tastes as good as it smells, but the seed is too big to be swallowed. Much pulp is lost as the young proboscidean works the seed out and over the edge of his mouth. Then he picks up another fruit, and another. Finally, he manages to crush a slippery seed between cusped molars. The mash is hastily rejected. The bitter toxins are the plant's way

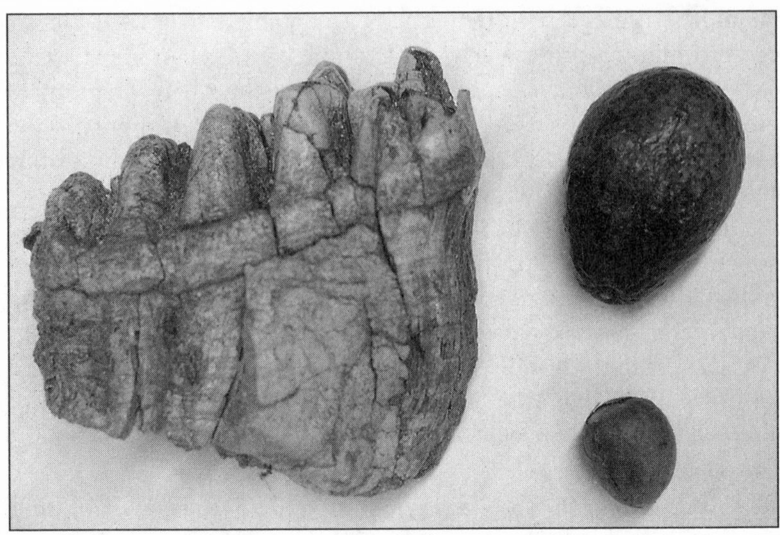

RELIC OF A GHOST. An avocado fruit and seed paired with the tooth of its missing partner in evolution, the gomphothere *Cuvieronius.*

of ensuring that dispersers do not become seed predators. Seeds are to be swallowed, not chewed.

Giving up, the young gomphothere nudges a cousin into play while the adults continue to eat. The avocados are soon gone, and the matriarch leads the herd to a forest clearing where browse is abundant. In a day or two, the gomphotheres will begin to deposit in fertile mounds the avocado seeds, along with many smaller seeds of other fruiting species ingested around the same time. Many of the seeds that survived the intestinal voyage will fall prey to seed-eating rodents or parrots, especially after dung beetles have carved the heap into fragments, rolling away the rich matrix to feed their young. Perhaps one seed will become a tree.

Ten years pass. A young avocado tree bears fruit for the first time. The gomphotheres discover it easily and add its location to clan memory. By no coincidence, the tree is near an ancient, well-worn path.

Fifteen thousand years pass. *Persea americana* still grows in Central American forests not yet turned into pasture. But the extent and density of the species do not match their former glory. A menagerie of small pulp thieves and seed predators raid the fallen fruits. Those who plunder the pulp leave behind seeds destined to compete with the parent. Seed predators may wait for the molecular transformations of germination that subdue the toxins, or they may gnaw into seeds to extract just the embryos.

Now that the migrators are gone, the sheer mass of fruit overwhelms the appetites of locally based thieves and predators. Molds attack the pulp of many overripe fruits. Fungal hyphae work their way into the seeds.

If a tree is very lucky, a jaguar may happen by. The avocado's oily flesh is attractive to this otherwise strict carnivore.[7] Because jaguar teeth are designed for tearing flesh—not grinding seeds—and because its gullet is adapted for swallowing great chunks of meat, a jaguar is a helpful seed disperser for avocado. But large carnivores were never abundant and are less so today. The avocado lineage may have been saved from extinction by the rare jaguar who took advantage of an easy meal, and perhaps by agouti rodents that gathered and buried avocado seeds just as squirrels gather and bury acorns.[8] The occasional pulp thief who scampers off with a fruit in its mouth, in order to strip off the pulp in a safer location, has surely helped the lineage as well. Nevertheless, the fruit of avocado was not shaped by millions of years of selection for these underabundant, ill-fitted, or fickle dispersal agents.

Nor was it shaped by the food preferences of bipedal apes, who invaded avocado territory just before the gomphotheres and ground sloths disappeared. Those apes are now doing a better job dispersing one species of the genus than any animal has done before. The dominant habitat for avocado today is in villages and orchards—and its range now wraps around the world.

Fortuitously, avocado was superbly built to attract the new mutualists. Nevertheless, it was not built to the specifications of apes. Rather, the fruit of the avocado (like that of mango, grapefruit, and pomegranate) was the plant kingdom's ingenious response to the pageant of beasts grown big throughout the Cenozoic and throughout the world. The beasts did not become giants in order to consume avocados. Their gigantism owes to millions of years of adaptive change to deter predators, to store energy for lean times, and to overpower rivals in mating jousts. In contrast, the avocado lineage did indeed evolve big-seeded, big fruits with the big beasts in mind. The bigger the seed, the better provisioned the embryo. Big-seeded plants have an advantage over small-seeded plants in mature forests, where sunlight penetrates to the ground only in patches and for maybe an hour or two each day. Big-seeded fruits of the tropical forests can grow for a year entirely on the energy sequestered in the seed. Perhaps during that pivotal year a tree will topple, allowing a shaft of light to penetrate. Or perhaps the seedling itself will reach a height where photosynthesis can begin in earnest.

Avocado's strategy for propagation made a great deal of sense throughout the long life of this lineage—until the present moment. Even after thirteen thousand years, avocado is clueless that the great mammals are gone. For the avocado, gomphotheres and ground sloths are still real possibilities. Pulp thieves like us reap the benefits. *Homo sapiens* will continue to mold the traits of the few species of genus *Persea* it prefers. Ultimately, however, wild breeds will devolve less grandiose fruits, or else follow their animal partners into extinction.

An avocado sitting in a bin at the grocery story is thus biology in a time warp. It is suited for a world that no longer exists. The fruit of the avocado is an ecological anachronism. Its missing partners are the ghosts of evolution.

Anachronisms Everywhere

Besides grocery stores, city streets and parking lots are excellent places to encounter anachronisms and the ghosts who loiter nearby. Honey locust *(Gleditsia triacanthos)* and ginkgo *(Ginkgo biloba)*, both abundant along the streets of New York City, bear anachronistic fruits. Urban landscapers did not, of course, plant these trees in order to entertain ghost watchers like me, but because of their hardiness to drought and pollution. Also, honey locust and ginkgo are attractive trees, and their roots are deep enough to leave the surrounding pavement unbuckled. Happy coincidence, therefore, has brought anachronisms to the streets of New York. And so, provisioned with the crucial scientific understanding and a bit of imagination, I can expect to enjoy a Cenozoic specter on my way to the subway and a far older ghost from the Mesozoic as I walk to the bookstore.

Especially in late fall and winter, I've caught glimpses of mammoths and mastodons along many of the sidewalks of New York—anywhere that banana-long pods dangle from barren honey locust branches. From a distance, these fruits of honey locust are strikingly similar to the *Cassia grandis* pods that prompted Dan Janzen's epiphany in Costa Rica. Ghost watching in the vicinity of a honey locust tree is possible year-round. The pods of one year are so weather-tight that they will linger on the ground into the next. There is, after all, no one here to eat them.

Ghosts of dinosaurs are easy to conjure in October and November wherever city landscapers planted ginkgo trees before they became acquainted with the fruit. In these older ginkgo plantings, where half the trees are female, even when I forget to look for the ghosts of dinosaurs my nose alerts me to their presence. Only a carrion eater could find the odor of fallen ginkgo fruit appealing. Before beginning this book, I wrongly blamed the alcoholic homeless for the vomitlike stench in Washington Square Park.

A PLEISTOCENE ANACHRONISM. Honey locust, *Gleditsia triacanthos,* bears fruit too big to be consumed by North American mammals that survived the Pleistocene extinctions. Though formerly restricted to eastern North America, honey locust is now a favored street tree in many urban areas throughout the world. Pictured here is a young honey locust with ripe pods on a New Mexico ranch.

Now that I am no longer blind to their presence, Cenozoic and Mesozoic ghosts have greeted me in a surprising number of places beyond New York City. Several years ago my sister planted three young ginkgo trees around her home east of Seattle. If at least one turns out to be female, the dinosaurs will surely pay a visit. My sister lives near Cenozoic ghosts too. I counted more than twenty honey locust trees while waiting for a bus at a park-n-ride near her home in Issaquah. Honey locust was the first tree I saw upon stepping out of an airport terminal earlier that summer in Rapid City, South Dakota. The next day, an hour's drive to the south, I came upon a dozen in the parking lot and along the walkway leading up to—yes, this is true—the Mammoth Site. This partially excavated hill at the edge of the little town of Hot Springs holds the largest aggregation of intact mammoth skeletons discovered in North America. Upon inquiring, I learned that the manager of the site was unaware that the grounds had been landscaped with a tree that so visibly remembers them. A few months later, honey locusts greeted me in Santa Fe and Los Alamos, New Mexico. Indeed, since I became aware of anachronisms and began scouting for them, I have yet to visit a city that lacks haunted trees.

Another haunted tree was until recently favored in rural landscaping: osage orange, *Maclura pomifera.* My first encounter with this North American native came after a talk I gave on ecological anachronisms at an ecospirituality conference in Bloomington, Indiana. During the lunch break, one of the participants led me to a magnificent old tree. Strange fruit—about the size, weight, and firmness of a softball and the color of a glow-green tennis ball—decorated the ground beneath the tree and out to about as far as one would roll. We collected several, returned to the conference hall, and placed one on the altar next to the photo of Earth from space and a bundle of ceremonial sage.

The participants had been quite taken with the idea that familiar plants inhabiting their region were ambassadors from deep history. These folks were well versed in the science of ecology, and, to a lesser extent, evolutionary biology, but they still looked to the stories of Native Americans for inspiration pertaining to the plants and animals around them. Coyote was a hero in many of these stories, but the great beasts of the Pleistocene had been forgotten.

"How do you spell *gomphothere*?" someone asked during the Q&A.

My next encounter with osage orange came a month later in an old Roman cemetery in Arles, France. Later that day I discovered that one such fruit had been moved more than two kilometers, probably by a juvenile of the tree's replacement dispersal species. The fruit had been deposited, uncrushed and sans dung, in a lousy place to sprout, however: a bench in the town plaza.

Up the Rhône River in the village of Sérignan, a single sample of the same anachronistic fruit was on display at the home of France's beloved naturalist of the nineteenth century, Henri Fabre. I came upon the fruit while strolling through the garden, after touring the room that had been his study and laboratory, its cabinets still full of carefully catalogued seashells, pinned insects, and fossil ammonites. The anachronism was on a stone table beneath a cedar tree. Where was its parent? The caretaker-guide seemed delighted by my question (spoken in halting French). He led me to a sapling with a metal label that read, "Orange d'osage, *Maclura pomifera.*" In his commentary, he seemed to be saying that because this tree was so young, he had to bring in a fruit from somewhere else.

Well that explains it, I guess.

Actually, it does. The fruit is so strange it makes collectors of us all.

Far and away the most surprising place I have encountered anachronisms has been in the little village where we spend the summer, surrounded by national forest lands in southwestern New Mexico. The ranch family here planted osage orange and honey locust for shade and orna-

An elephant fruit in a land without elephants. The bright green fruit of osage orange, *Maclura pomifera,* was shaped by the now extinct megafauna of North America.

ment. Their livestock trample but will not eat the spheres of osage orange, which ripen on the ground from green to yellow after several weeks of frosts. The horses ignore the honey locust pods as well, though the mules do take a few that have fallen. Only the little burro plucks the pods before they fall.

Equids very similar to today's horses and burros evolved in North America in the company of honey locust during the past thirty million years. Ancient equids unaccustomed to bales of hay would surely have been eager to serve the reproductive interests of a tree that offered a nutritious meal. As well, our kind has been assiduously breeding honey locust into a degenerate form. We love the lacy foliage but deem the pods a nuisance. Varieties sold today for landscaping bear smaller and less pulpy pods than do wild populations. The newest cultivars are clones of male stock. Wherever these clones are planted, ghosts will not appear.

Another anachronism grows near our New Mexico home, exclusively in the wild. Along the river and the road where the soil has been disturbed, I sometimes encounter the desert gourd, *Cucurbita foetidissima*. For years, I wondered why the fruits would lie rotted on the ground even as the next year's crop was ripening. I certainly would not want to eat these foul-smelling, bitter-tasting gourds even when fresh, but I wondered why the plant wasted so much energy growing pulp that nobody else would eat either. Now I know the answer: ecological anachronisms are everywhere.

A Brief History of the Scientific Search for Ghosts

Dan Janzen and Paul Martin did not lack predecessors. Previous work, however, aimed to solve isolated ecological puzzles, not to introduce a new evolutionary concept. Whereas "Neotropical Anachronisms" pertained to an entire hemisphere, two earlier papers dealt with odd or inexplicable plant traits on particular islands.

In 1977 R. M. Greenwood and I. A. E. Atkinson published "Evolution of Divaricating Plants in New Zealand in Relation to Moa Browsing" in the *Proceedings of the New Zealand Ecological Society.* Species of ostrichlike moa are the ghosts; the divaricating growth form is the anachronism. A branch grows outward, an offshoot angles widely to the side, and the next may head back toward the plant's interior. An impenetrable mesh of interwoven twigs is the result. This strategy of growth is excellent defense against browsers. It is, however, expensive. A lot of structural material is required to support foliage that is not particularly well situated to capture solar energy.

Ten percent of the woody flora native to New Zealand have a divaricating pattern of growth; elsewhere in the world such plants are rare. It is safe to assume that a lineage expressing this growth habit must have had good reason to do so. Because so many New Zealand plants of diverse heritage evolved divaricate stems independently, biologists would be remiss not to look for adaptive explanations. Greenwood and Atkinson argue that the shaping influences cannot be detected today because they are extinct. Millions of years of exposure to the browsing of moas encouraged many plants (particularly those whose leaves are not chemically defended) to evolve into thickets. Spines and thorns may be fine armaments against the sensitive snouts of mammals, but only a thicket could frustrate the tough bills of New Zealand's giant vegetarian birds. Seven centuries have passed since the dozen species of moa vanished—too little time for the plants to have altered their forms.

Also in 1977 Stanley Temple published a two-page paper in a journal with a far wider readership. "Plant-animal mutualism: coevolution with dodo leads to near extinction of plant" appeared in the same journal Janzen and Martin would publish in five years later: *Science.* Like the moa, the dodo was sent into extinction because seafaring humans who colonized its island home found this flightless bird edible and easy game. Temple suggested that the dodo had played a vital role in dispersing a particular tree found only on the island of Mauritius in the Indian Ocean, which, in turn, was the only place where the dodo had lived. He pegged the demise of the dodo three hundred years ago as the reason the magnificent tambalacoque tree had drifted toward extinction.

Both the moa and dodo hypotheses generated debate. As we shall see, their core contentions have withstood the test of time. Divaricating plants of New Zealand and the fruit of the tambalacoque tree of Mauritius are bona fide anachronisms.

Though I began this project as a science writer aiming to produce a wide-ranging review of a vibrant area of research, the quest for botanical anachronisms is so new and the uninterpreted data so rich that I must offer some fresh ideas of my own. Most important is the suggestion that anachronism is best understood not as a binary concept, an either-or. Rather, the fruits and armaments produced by plants are anachronistic to varying degrees. Some are extremely anachronistic, but most are only moderately ill fitted for the animals with whom they now must interact.

Dan Janzen goes even farther. He maintains that, strictly speaking, all traits are anachronistic, whether we can discern the lagging features or not. "Every organism is a collection of anachronistic traits," he told me. "Put another way, each trait is to some degree anachronistic. It is absurd to think that the selective pressures that generated some particular trait are present right now and here today. The closer you get to simple organisms, the greater the chance that its traits are less anachronistic and more set by current selective pressures. But even with viruses there are all sorts of phylogenetic inertia, historical accidents, anachronisms—whatever you want to call them."[9]

Perhaps the most helpful definition of *anachronism* that Janzen has offered in print is the one that appears in a private letter to Gary Paul Nabhan, dated 25 June 1987. "If a tree evolved a trait in response to some pressure," Janzen wrote, "and that trait is around later on despite that pressure having been lifted, then that trait is an anachronism."

Anachronisms may not only be everywhere but in every body. They are likely in humans more than most because we have cultural fixes for physical (and to some extent psychological) misalignments. And we, more than most, are able to pass on funky old genes long after their usefulness has dwindled. I know by hard experience that my left wrist was not designed for a keyboard; it is bound in a brace as I type. I know, too, that my back was not engineered for hours of sitting in one spot or my eyes for finding prey on a lighted screen.

I do not, however, marvel at these dysfunctions. Yet I do marvel at the fruits of honey locust, osage orange, ginkgo, avocado, papaya, and others whose stories you will encounter in this book. What is the difference between one set of anachronisms and the other?

Here I must depart from the perspective of Dan Janzen and turn to that of John Byers, a zoologist at the University of Idaho. In 1997 Byers reported a stunning animal anachronism—although he used the term *relict behavior* to characterize it. The relict behavior Byers discovered is the excessive speed of the American pronghorn.[10] The sole surviving member of the family Antilocapridae, this deerlike herbivore (with hooves and horns more like a goat's) lives on the plains of North America. The pronghorn is exceptionally fast, sprinting up to sixty miles per hour. Yet no predator in the Western Hemisphere comes even close to matching that speed. Why, then, does the pronghorn run so fast? Byers believes the pronghorn is still fleeing predators who were faster and more formidable than coyotes and wolves. A big hyena, a long-legged bear, and a cheetah and a lion nearly identical to their African kin had pressed the pronghorn to run at its astonishing pace. These predators vanished thirteen thousand years ago, along with the gomphotheres, mammoths, and ground sloths.

Byers introduced more than a terrific example of an anachronism. He offered an easy way to identify traits that deserve scrutiny: they are overbuilt. It is the overbuilt quality of biological structures, physiologies, or behaviors that nudges us to look for missing partners—organisms (or organism types) once vitally important in shaping traits of another lineage but alive no more.[11]

Pronghorn are overbuilt for running on the post-Pleistocene plains of North America. Should the plains stay as they are now, the pronghorn might—in a few tens of thousands of years—reroute some of that capacity for horizontal movement into the vertical. Unlike deer, pronghorn are poor jumpers. They had no need to jump on the Pleistocene plains. Jumping would be a big help now, however, because their habitat is crisscrossed by barbed-wire fences. An inability to jump thus puts the pronghorn at a severe disadvantage—a life-threatening disadvantage when a migratory route or a water source is blocked or when a fawn wriggles under a fence but cannot return to its mother.[12]

Overbuilt applies just as well to avocado, papaya, honey locust, osage orange, and desert gourd. The fruits are either too big for the gapes of currently available mouths or the pulp is laced with toxins that no stomach will tolerate today. All such anachronisms are thus haunted by ghosts. In contrast, my wrists, my back, my eyes are not missing any identifiable partners in evolution, and it would be a stretch to consider them overbuilt for current needs.

The anachronisms featured in this book are missing their longtime partners in ecological interactions, and they are stunningly overbuilt. In the case of the pronghorn, being overbuilt may not have deleterious consequences that would impede survival of the lineage. In contrast, the avocado is overbuilt in a way that not only wastes energy but seriously hinders dispersal of its seed. A wild avocado is Ginger Rogers without her Fred Astaire. Jerry Lewis is now her only suitor, and he keeps tripping over his own feet.

Bringing Evolution into Ecology

"Living organisms are beautifully built to survive and reproduce in their environments. Or that is what Darwinians say. But actually it isn't quite right. They are beautifully built for survival in their *ancestors'* environments."

Richard Dawkins made that statement to a gathering in London in 1998.[13] As we shall see, exemplars of his message include plants with big fruits or ferocious thorns in tropical and temperate America. These botanical features—all made anachronistic by the loss of once influential partners—are a subset of Dawkins's broader mix of examples. They are all examples of biology out of sync.

"The genes that survive down the generations," Dawkins continued, "add up to a description of what it took to survive back then. And that is tantamount to saying that modern DNA is a coded description of the environments in which ancestors survived."

The concept launched by Janzen and Martin in the field of ecology is easily subsumed within the worldview of evolutionary biology. Indeed, it has a long history within that discipline, beginning with Charles Darwin. Among evolutionary biologists it is referred to simply as time lags. In a book published the same year that Janzen and Martin showed the relevance of the anachronism concept in their own field, Richard Dawkins wrote (without fanfare), "Since modern man has drastically changed the environment of many animals and plants over a time-scale that is negligible by ordinary evolutionary standards, we can expect to see anachronistic adaptations rather often."[14]

Dawkins used the word *anachronistic* in precisely the way Janzen did. The synchronicity did not, however, mean that something spooky was going on in the universe. The gods were not insinuating a new revelation into human consciousness. If one could take the time to look, *anachronistic* and *anachronism* would surely appear in many older writings of evolutionary biologists, and perhaps some ecologists too.[15] Why, then, the fuss?

Why was Janzen and Martin's take on the matter of time lags deemed sufficiently innovative for publication in the top scientific journal? And why did that paper then inspire fellow ecologists?

Ecologists are not rigorously trained in evolution—at least they were not in Dan Janzen's generation, and Janzen told me that he had grown up "intellectually ignoring the fossil record."[16] Theoretical evolutionary biology was not a research subject that ecologists paid much attention to. There was more than enough to keep track of in their own, newly burgeoning field.

Ecology is the science of interactions taking place today. Paleoecology is the science of interactions in bygone times. Evolution is the thread that connects the two realms. A new interdiscipline called *evolutionary ecology* evolved between the cracks in the academy about the same time Janzen embarked on this new path. The Janzen and Martin 1982 paper is thus a classic contribution to this still youthful interdiscipline.

All of Dawkins's examples in his 1982 treatment of "anachronistic adaptations" are here-and-there curiosities. He described moths genetically programmed to rise toward the stars now annihilating themselves in candle flames. Another example: "The hedgehog antipredator response of rolling up into a ball is sadly inadequate against motor cars." In contrast, there is nothing inadequate about an avocado. Avocado is too much of a good thing: overbuilt, overdone, and overcommitted for a function that still very much pertains. More, an avocado is not an isolated curiosity. It is one of a great basket of fruits all bereft of animal partners that disappeared thirteen thousand years ago and that can be found variously throughout an entire hemisphere. Humans are likely the ultimate cause of this disjunction in adaptations, as we are for hedgehog roadkill. But the human intervention that concerns avocado happened well before the automobile.

Evolutionary biologists have long known that anachronisms are reality. Ecologists have shown that anachronisms are a stunning component of reality. The plant realm lingers in a lost world.

On the other side of the research divide in evolutionary biology is Stephen Jay Gould. His anti-adaptationist views should resonate with the finding that avocado is not adapted for anything around today. To insist that avocado is adapted to squirrels or monkeys is to force a round peg into a square hole. But Janzen and Martin's insistence that adaptation, albeit adaptation drawn out of the past, is the ultimate answer to the ecological puzzles they address is not anti-adaptationist.

Gould's view of contingency does, however, meld with the anachronism idea. Avocado did not deserve to be left out to rot. The lineage did not do anything foolish that should have put it at risk of range retrenchment and, ultimately, extinction. Avocado is a fine form of fruit, and megafauna were dependable mutualists for tens of millions of years. But things happen that cannot be anticipated. Even if change could be anticipated, natural selection does not anticipate.

Gould and coauthors coined two new terms in evolutionary biology—*spandrel* in 1979 and *exaptation* in 1982—that applied to non- or extra-adaptive traits.[17] The published papers launching each newly named concept included a discussion of the extent to which Charles Darwin had anticipated such ideas in *Origin of Species*. None of the participants in the anachronism research program has yet made this journey back to the source. Did Darwin anticipate anachronisms the way he seems to have anticipated spandrels and exaptations?

Several passages in *Origin of Species* are relevant. Darwin wrote, "But by far the most important consideration is that the chief part of the organisation of every being is simply due to inheritance; and consequently, though each being assuredly is well fitted for its place in nature, many structures now have no direct relation to the habits of life of each species."[18] The examples Darwin provides—the webbed feet of the so-called upland goose and the homologous bones in the forelimbs of horses, bats, and seals—don't have anything to do with missing partners and offer little to illustrate the concept of overbuilt parts. They are, rather, examples of what has come to be called *phylogenetic inertia.* Phylogenetic inertia (also known as stasis) is biological resistance to change, and for any number of reasons. It is about getting by with what is good enough; if it ain't (totally) broke, don't fix it.

Phylogenetic inertia is crucial for understanding why divaricate plants in New Zealand haven't yet noticed the loss of the moa, why the avocado hasn't caught on that the gomphotheres are gone, why the pronghorn refuses to slow down. But it is too broad a concept to distinguish anachronisms haunted by ghosts. For example, phylogenetic inertia (coupled with exaptation) in part explains why giant ground sloths walked on the sides of their feet. Phylogenetic inertia (more precisely, biological constraint) is the reason vertebrates, unlike insects, cannot sprout wings from their backs. The concept also applies to the human appendix and to the vestigial pelvic bones or hind limbs in some whales and snakes. Darwin grouped these latter examples under the term *rudimentary organ.*

Rudimentary organ cannot, however, describe the vigor of an avocado fruit. Ecological anachronisms are anything but rudimentary. They are not shriveled vestiges of former functions that no longer apply. Rather, they are outrageously overbuilt for functions that still very much matter. Those functions—be they attraction of seed dispersers, repulsion of pulp thieves, or escape from enemies—could be served more economically, and in many cases more successfully, if the overbuilt trait were reined in.

Darwin continues, "Hence every detail of structure in every living creature (making some little allowance for the direct action of physical conditions) may be viewed, either as having been of special use to some ancestral form, or as being now of special use to the descendants of this form—either directly, or indirectly through the complex laws of growth."

A long paragraph of examples and argument follows, concluding with: "Natural selection will never produce in a being anything injurious to itself, for natural selection acts solely by and for the good of each. No organ will be formed, as Paley has remarked, for the purpose of causing pain or for doing an injury to its possessor. If a fair balance be struck between the good and evil caused by each part, each will be found on the whole advantageous. After the lapse of time, under changing conditions of life, if any part comes to be injurious, it will be modified; or if it be not so, the being will become extinct, as myriads have become extinct."

Yes, indeed, Darwin does anachronisms—though he does not highlight the haunted variety, nor does he suggest a name for the idea. The only issue is what the master might have intended by the phrase "after the lapse of time." Is seven hundred years too brief for the divaricate plants of New Zealand to respond to a change in herbivory? Should thirteen thousand years have been enough to get honey locust to think about wooing someone of lesser stature than an elephant? What about sixty-five million years: Would Darwin have expected the devolution or demise of ginkgo over such a long time?

This is how Darwin addresses time lag: "Though Nature grants long periods of time for the work of Natural Selection, she does not grant an indefinite period; for as all organic beings are striving, it may be said, to seize on each place in the economy of nature, if any one species does not become modified and improved in a corresponding degree with its competitors, it will soon be exterminated."[19] (For the sixth edition of *Origin*, Darwin—significantly, in our context—eliminated the word "soon.")

Darwin presages John Byers's characterization of a variety of relict behaviors in animals as overbuilt. It seems that Darwin, too, would have been puzzled by the excessive size of avocado seed in today's neotropical forests and by the excessive speed of the pronghorn on North American plains. These traits call for uncommon scrutiny. Darwin writes, "Natural selection is continually trying to economize every part of the organization. If under changed conditions of life a structure, before useful, becomes less useful, any diminution, however slight, in its development will be seized on by Natural Selection, for it will profit the individual not to have its nutriment wasted in building up a useless structure."[20]

In the chapter summary, Darwin restates this point: "Any part or organ developed to an extraordinary size or in an extraordinary manner, in comparison with the same part or organ in the allied species, must have gone through an extraordinary amount of modification since the genus arose; and thus we can understand why it should often still be variable in a much higher degree than other parts; for variation is a long-continued and slow process, and Natural Selection will in such cases not as yet have had time to overcome the tendency to further variability and to reversion to a less modified state. But when a species with any extraordinarily developed organ has become the parent of many modified descendants—which on my view must be a very slow process, requiring a long lapse of time—in this case, Natural Selection may readily have succeeded in giving a fixed character to the organ, in however extraordinary a manner it may have been developed."

Perhaps large size and exceptional pulpiness are characters so "fixed" in avocado and honey locust that thirteen thousand years has been insufficient time for "diminution" of these traits. Perhaps without our own intervention in their behalf, population shrinkage would proceed to the point of "extermination"—before genetic adjustments could "economise" the fruit form. As we shall see, honey locust may in fact have adjusted; the fruit of offshoot populations may indeed have devolved. And if any plant is capable of persisting for sixty-five million years with a severely debilitated dispersal system, it would surely be the obstinate, nearly invincible ginkgo.

Darwin was keenly aware of extinction, but not of mass extinction. And Darwin didn't know that, for sad reasons indeed, he and we inhabit a slot of evolutionary time that is oozing with freshly minted ghosts. We can experience ghosts because we live in a mass extinction of our own making. The preamble to this now catastrophic episode for earth life thundered in with what Paul Martin and David Burney call "extinction of the massive."[21]

Extinction of the massive began about fifty thousand years ago in Europe and thirty thousand years ago in Australia. It devastated the Western Hemisphere thirteen thousand years ago and ravaged Madagascar two thousand and the islands of New Zealand less than one thousand years ago. And lo, those dates are also the times when the human presence in each of these lands can first be detected. Martin concludes that this is more than mere correlation. His "Pleistocene Overkill" theory posits cause.[22] We spear-toting, fire-brandishing humans overhunted the great herbivores into oblivion. The small and fragmented populations of slow-reproducing survivors were doomed—whether or not we managed to put a point through the heart of each or to drive every herd over a cliff. In the trophic cascade that followed, the carnivores and scavengers dependent upon the megaherbivores vanished too.

Most of this book surveys ecological anachronisms that my generation and yours bear no responsibility for. Chapter 9 looks at those on the threshold: anachronisms in the making.

Adding Magic to a Walk

This book describes a scientific research program in its infancy. To report on and thus advance the science is not, however, the only reason I chose to spend two years on this topic. I also hope to show that the wondrous world is even more wondrous than we imagine. Some days I can barely control my urge to stop the next passerby on Laguardia Street and demand, "Hey, look! There's a ghost behind that tree!"

This research and writing venture came about primarily because Paul Martin likes to proselytize too. I first encountered the ghosts concept in 1992 in a popular journal at the radical end of the conservation movement: *Wild Earth*. It contained an essay by Martin titled "The Last Entire Earth." He wrote, "In the shadows along the trail, I keep an eye out for ghosts, the beasts of the Ice Age. What is the purpose of the thorns on the mesquites in my backyard in Tucson? Why do they and honey locusts have sugary pods so attractive to livestock? Whose foot is devil's claw intended to intercept? Such musings add magic to a walk and may help to liberate us from tunnel vision, the hubris of the present, the misleading notion that nature is self-evident."

Evolutionary ghost stories are necessarily speculative. At their core, however, is the sound contention that when structures or behaviors are puzzling, one should look for adaptive explanations in the past before giving up the ghost.

Attempting to wrest adaptive explanations out of present circumstances can lead to excesses of yarn-spinning far worse than anything I, or the scientists included here, shall offer. Consider this adaptive explanation for the honey locust pod, published in 1917: "The purple pods cling and rattle in the wind long after the yellow leaves have fallen. One by one, they are torn off, their S-curves tempting every vagrant breeze to give them a lift. On the crusty surface of snowbanks and icy ponds, they are whirled along, and finally lodge, to rot and liberate the seeds. It takes much soaking to prepare the adamantine seeds for sprouting. The planter scalds his seed to hasten the process. Nature soaks, freezes, thaws them, and thus the range of the honey locust is extended."[23]

Okay, maybe newly weaned gomphotheres like the one in my story didn't spit out the first avocado seeds they managed to crush. Maybe they ground those seeds to bits and swallowed—then sickened and threw up.

(A South American folk recipe for rat poison is cheese or lard spiked with ground avocado seed.) Gomphotheres that reacted to the bitterness of avocado seed (it tastes repulsively bitter to me) by learning to eat the whole fruit properly were better able to produce progeny—and progeny with the same knack. One way or another, avocado and gomphotheres became mutualists. And if not exactly the gomphotheres, then maybe the giant ground sloths or notoungulates like toxodon forged the extravagance of avocado fruits.

Adaptive just-so stories *are* sometimes a more prudent response to an evolutionary puzzle than to throw up one's hands and sigh, "It just happens to be that way, I guess," or "That's the way the genus (or family) has always built a fruit: it's phylogenetic inertia." Sometimes it is prudent to hypothesize that the trait in question is, at least to some degree, anachronistic, and then to search for missing partners.[24]

Evolutionary ghost stories offer the best causal explanations for some botanical and zoological puzzles. They also entertain our story-loving minds. Because of the role we humans play in all stories that feature Pleistocene or more recent ghosts, the tales are poignant too. Perhaps a modern-day Kipling or Aesop will render them as fables for the generations of children who will grow up in a world where the survival of tens of thousands of species will depend on whether humans learn to care.

ANACHRONISTIC FRUITS OF THE NEW WORLD TROPICS. The tough-rinded spheres of two species of *Crescentia* (*C. cujete* left; *C. alata* center), along with the curled pods of three species of *Enterolobium* (*E. cyclocarpum* lower right), are food for large Pleistocene mammals—ranch horses—that have been restored to their native range.

2

Ecological Anachronisms and Their Missing Partners

PAUL MARTIN AND I ARE BROWSING THROUGH FOLDERS OF DRIED plants in the herbarium of the University of Arizona, where he is now professor emeritus. Earlier that morning he gave me stem and pod samples of several kinds of native and nonnative mesquites and acacias growing around his home. I have come to Tucson to talk with Martin about anachronisms, such as the seed pods of the ubiquitous legume trees of the Sonoran Desert. The trip to the herbarium is a bonus.

The University of Arizona Herbarium, tucked into a basement corner of a building on the Tucson campus, has a cozy, laid-back feel—quite unlike the vast corridors of the New York Botanical Garden Herbarium, where some of the photos you will encounter in this book were taken. Both storehouses of dried plants are nevertheless suffused with the same faint aroma. Like Pavlov's dog, I have grown to savor the scent even more for the attractions that come with it. In both herbaria, plant specimens are filed in tall metal cabinets organized by taxonomic family. Within each cabinet, oversize manila folders containing sheets of pressed plants are stacked alphabetically by genus. Color cues signify broad geographic differences in collecting sites. The herbarium in New York, one of the world's great repositories, uses seven colors. Here in Tucson, three are sufficient, owing to the regional focus. Yellow folders indicate Arizona, green is for Mexico, and pink for everywhere else. Everywhere else is usually Texas but can be as distant as Africa.

Fortunately, some of the pink folders contain Old World specimens to compare with the dominant yellow and green folders of the two expansive subfamilies of the legume, or pea, family: Mimosoideae and Caesalpinioideae. Some legume pods in both subfamilies are slender, with paper-thin husks that dehisce explosively upon ripening, propelling small seeds a meter or so away from the parent plant. Other pods (Australian *Acacias* especially) open without spectacle, relying instead on ants enticed by a fatty packet attached to the seed coat to give the tiny propagules a degree of mobility. Still others enlist the services of wind. None of these dispersal strategies interests Paul Martin and me right now. Only the large, woody, indehiscent pods that retain their seeds long after ripening attract our attention. Large, woody, indehiscent pods in pink files from Africa are known to be dispersed by elephants and giraffes, and also by rhinos and lesser beasts that take whatever may fall after the primary dispersal agents have conducted their aerial harvests. Because such dispersers are not yet extinct in the native habitats of the pod-bearing plants, these fruits are not yet anachronisms.

Some pods in the yellow and green files of Arizona and Mexico are strikingly similar to the African pods. These were, until recently, assumed to be scattered by rodents or flowing water or (when all other explanations fail) gravity. But here in this herbarium, the existence of anachronisms seems unassailable. Surely the large, woody, indehiscent pods endemic to North America evolved to attract elephants too: the gomphotheres, mastodons, and mammoths. Some of these fruits could probably claim affiliations with American species of rhinoceros as well, which vanished from this hemisphere just five million years ago, long after the fundamental fruit forms were established. Perhaps some retain features they initially developed to woo the giant herbivorous brontotheres (also known as titanotheres), which preceded the beasts of the Pleistocene and Pliocene by tens of millions of years.

Pods now native to Arizona and Mexico but originally from South America almost surely were shaped not by gomphotheres but by giant ground sloths, camel-like liptoterns, and chunky notoungulates—all extinct a mere thirteen thousand years ago. These beasts evolved and lived only on that continent while South America was still isolated from the rest of the world, separated from its neighbor to the north by a substantial seaway. From the plants' perspective, extinct gomphotheres and the few large (but far from elephantine) fauna now native to South America—deer, tapir, peccary, llama, bear—were recent arrivals. These mammals (and many smaller ones too) migrated to South America only during the past three million years, after the Isthmus of Panama rose from the sea.

The ghosts of evolution thus haunt the Mimosoideae and Caesalpinioideae cabinets of the Arizona herbarium. To understand the dispersal adaptations of the forty-some species of small trees in the mesquite genus, *Prosopsis* (subfamily Mimosoideae), or that of the much thicker and tougher pods of *Hymenaea courbaril* (subfamily Caesalpinioideae), one must conjure the ghosts of strange, extinct mammals in the land where these plants originated: South America.

There are human ghosts here as well. Howard Scott Gentry, who collected many of the pressed plants we examined, is one of Paul Martin's favorites. It is standard practice for the collector or herbarium staff to affix labels onto the specimen sheets that provide the collector's name, along with the date and precise location where the specimen was taken. Descriptive details are also sometimes offered, and Gentry's descriptions are more than occasionally artful. A specimen of *Acacia brandegeana,* collected on the Baja Peninsula in 1939, bears this legend: "A tough polypodial shrub bent from the hurrying press of flood waters."

Paul Martin's name appears on a number of sheets, but his descriptions are perfunctory.

Delonix elata
21 Nov 1965
collected by Paul S. Martin
Magadi, Kenya; elevation 1975 feet

"What were you doing in Africa?" I ask.

"I was there to visit Louis Leakey. I was working on my overkill paper then and wanted to learn about the extinct suids, baboons, and sabertooth cats in Africa."

"Africa and Pleistocene Overkill" by Paul S. Martin was published in *Nature* one year later. This paper officially launched the idea that rapid colonization of the New World by humans ("fluted-point hunters") who had migrated across the Bering Land Bridge less than fifteen thousand years ago was the reason that mammoths, mastodons, ground sloths, glyptodonts, horses, camels, liptoterns, notoungulates, and other large mammals went extinct in the Western Hemisphere. It was this paper that made Paul Martin the intellectual partner of choice when Dan Janzen began to ponder what he called "the riddle of the rotting fruit."

Riddle of the Rotting Fruit

Dan Janzen chose the dry forests of western Costa Rica for his ecological fieldwork beginning in 1963, and that association continues to this day. By the time he commenced work on anachronisms, he had already published important papers on an astonishingly wide range of ecological topics. His leadership in conservation ecology was established with his early warning that the current extinction crisis pertained not just to the loss of species but to "the extinction of ecological interactions."[1] Later he would pioneer a pragmatic approach to conservation in his successful advocacy for establishment of a new national park—today, the Area de Conservación Guanacaste—in the very region where he conducted his fieldwork.[2] The Guanacaste conservation area was unusual in that some human uses that had become traditional in the area were not only to be tolerated under the new management status but outright encouraged for their ecological benefits. The horses and cattle associated with ranching were welcomed as dispersers of native trees that had lost Pleistocene partners. Janzen viewed these animals as vital to the reforestation of centuries-old pastures.

In their 1982 paper Janzen and Martin wrote, "We propose an answer to the riddle of why certain trees produce far more edible fruits than their current dispersal agents will remove, produce fruits that are not eaten by contemporary dispersal agents, bear fruits that resemble those eaten by African megafauna, and bear fruits avidly eaten by introduced livestock. . . . We are confronted with a number of puzzling fruit and seed traits whose mystery disappears when interpreted in light of the extinct Pleistocene megafauna."

They continued, "We are considering a portion of what happened when roughly three-quarters of all the species and individuals of large mammals were suddenly removed from a dry tropical region and its adjacent rainforests. The present-day analogy is a tropical, forested African habitat stripped of its elephants, rhinoceroses, zebras, elands, bush pigs, and other large herbivores and left alone for 10,000 years."

Janzen and Martin proposed two ways to test the anachronism idea: by comparison and by simulation. For the first test, they suggested "comparing the array of fruits eaten and seeds dispersed by large mammals in Africa and Asia with the fruits of tropical America on the one hand and with the fruits of New Guinea or tropical Australia on the other; the latter two tropical land masses have never had a mammalian fauna that would select for a well-developed megafaunal dispersal syndrome."

Not so, corrected ecologist Thane K. Pratt, in a response published in *Science*'s letters column.[3] Rather, "Australia and its land-bridge island, New Guinea, may provide support for their theory through an extinct megafauna and extant flora that still show adaptations of past coevolution." Pleistocene marsupial mammals in Australia included giant kangaroos and wombats, along with a tanklike animal *(Diprotodon)* that bore little resemblance to anything alive today. Pratt listed seven fleshy fruits native to New Guinea that may have evolved to attract these giants. The fossil record is less well known in the richly forested land of New Guinea than it is in Australia, he wrote. Nevertheless, bones of the same kinds of giant herbivores have been unearthed, though from sediments deposited during the geological epoch just prior to the Pleistocene.

Janzen and Martin acknowledged Pratt's correction in print in the same issue of *Science*. Unfortunately, a correction does not enjoy the fanfare of the text it modifies. No one seems to have picked up the banner and begun a thorough search for candidate anachronisms in Australia or New Guinea, although Australian ecologists have occasionally mentioned prospects.[4] Even so, in a cursory examination of published reports on the dispersal adaptations of Australia's many species of genus *Acacia*, I did come upon intriguing candidates.

About two-thirds of the world's 1,200 species of *Acacia* (family Leguminosae, subfamily Mimosoideae) are native to Australia. Most occur nowhere else, and almost all seem to have fruit traits clearly aimed at bird or ant dispersal. But almost all is by no means all. In a book chapter on Australian acacias, Dennis J. O'Dowd and A. Malcolm Gill present data on the fruits of ninety-two *Acacia* species.[5] Among these, I noticed seven whose large size precluded ant dispersal yet whose lack of a fleshy sac around each seed (an aril) was incompatible with bird dispersal. Also anomalous was the description of a pod in another paper that examined the dispersal strategies of twenty species of Australian acacia.[6] There the authors concluded that the "relatively large seed size" of *A. ramulosa* meant this species was not dispersed by ants. But the low energy-to-water ratio of its pod contents made it unattractive to birds as well. Now consider the ecological description: "*Acacia ramulosa* drops its massive legumes directly beneath the shrub, where we found them in abundance and unopened months after the fruiting season had concluded. . . . The aggregation of many weathered but unopened legumes beneath parent plants is also consistent with the apparent lack of adaptations for dispersal by ants and birds." The authors did not broach an adaptive interpretation.

This sounds to me like "the riddle of the rotting fruit," translocated from tropical America to Australia. Although the authors cited the Janzen and Martin anachronism paper, they did not use it to explain the anomalous legume. Perhaps they were reluctant to turn an anomalous result into a positive finding (discovery of a bona fide Australian anachronism) because they, like many others, seem to have missed the correction posted by Janzen and Martin.

Before anyone gets overly excited about *A. ramulosa* and the other anomalous pods, it will be crucial to confirm that these species have not themselves been translocated. For example, *A. nilotica* and *A. farnesiana* are now common in parts of Australia, but they are recent invaders from Africa and Central America, respectively. Their abundance in Australia owes to the pods' attractiveness to cattle, another translocated species. Should someone eventually conclude that fruits with anachronistic traits are in fact indigenous to Australia, the age of known Cenozoic anachronisms will have more than doubled. Australia's Pleistocene megafauna vanished not thirteen but thirty or forty thousand years ago.

What about the second test of the anachronism idea that Janzen and Martin proposed, the test of simulation? In their 1982 paper they wrote, "We can also test our hypothesis by reintroducing Pleistocene mammals such as horses to the neotropics and observing their response to the fruits and the response of the plant populations to the mammals." With about as much dry humor as one can sneak into *Science*, the authors noted that "the experiment has been running for 400 years." All that was needed was for scientists to make the relevant observations. Dan Janzen would lead the way.

Ranch Horses as Pleistocene Megafauna

It was a legume pod of the American tropics, *Cassia grandis*, that prompted Dan Janzen in 1977 to ponder "the riddle of the rotting fruit." A big-seeded palm fruit, *Scheelea rostrata*, was Janzen's choice for an imaginative story of gomphothere-fruit interaction that dramatized the science in his 1982 paper with Paul Martin. Janzen had written on the ecology of both *Cassia grandis* and *Scheelea rostrata* several years before the anachronism idea forced him to walk away from some of his earlier conclusions. Coming upon these and other early Janzen papers, I was struck by how much he had to strain to offer adaptive explanations for large Costa Rican fruits.[7] He sometimes mentions cattle as fruit eaters in these early papers but quickly dismisses them. That nonnative livestock could have anything to teach us about the evolutionary ecology of native fruits was not conceivable at that time.

Janzen tested four additional fruits of Costa Rican dry forests—*Guazuma ulmifolia, Spondias mombin, Crescentia alata,* and *Enterolobium cyclocarpum*—for megafaunal suitability during the several years that he and Martin were fleshing out the anachronism idea and readying their paper for publication. The results of these tests strongly support the anachronism hypothesis.[8] An important component of each study was how well the fruits are dispersed by horses that graze the countryside today. Horses, like cattle, can serve as convenient proxies for the set of large mammals that inhabited the Americas during and prior to the Pleistocene. Horses, in fact, are more than proxies: they *are* Pleistocene megafauna. Janzen regards horses in Costa Rica today as "a Spanish gift from the Pleistocene—invented here, then extinguished here by people, but surviving in the Old World 'refuge' and brought back here by people."[9]

Horse fossils of Pleistocene age have been found in Central America. Recall that Paul Martin had suggested "some tropical horse" in the preliminary list of Pleistocene dispersal agents that he sent to Janzen to confirm their collaboration. Yet the realization that ranch horses *(Equus caballus)* were not just proxies for extinct beasts *(Equus fraternus)* but the real thing began with an encounter Janzen had with "a pile of horse dung in the middle of the road, with a *Crescentia* seed on top." Later, while hand-feeding *Crescentia* fruits to horses for the dispersal tests, he realized that "I had an escapee from the Pleistocene extinction in my front yard."[10]

Crescentia alata (hereafter, crescentia) is a small tree of family Bignoniaceae. It is known locally in Costa Rica as jicaro and in Mexico as the calabash or gourd tree. Crescentia is native to dry forests and savanna of the American tropics, but it is patchily distributed. As Janzen observed, the tree "would not look out of place in Nairobi National Park in Kenya." The spherical fruits range in size from that of a small orange to much larger than a grapefruit. Like oranges and grapefruits, jicaro fruits have a rind. The fruits of a sibling species, *Crescentia cujete,* which is also found in the neotropics, can be as big as a soccer ball. The rinds of both species are hard and exceptionally tough to crack. Only horses are able to accomplish this feat—no "native" species can do it. The largest native herbivore, baird's tapir, cannot open its mouth wide enough to position incisors in a way to penetrate the husk.[11] Horses must use their hooves to crack open the spheres of the biggest fruits of *C. cujete.* Without these recruits from ranches, therefore, the fruits rot where they fall. Janzen calculated that it takes on average two hundred kilograms of pressure for horses to crack jicaro rinds.[12] Inside are several hundred stiff seeds, more rubbery than hard. The seeds are surrounded by "a slimy black mass that is quite sweet," Janzen reports, despite its fetid odor. Horses break the rinds between their front teeth, dropping the pieces to scoop out pulp with lower incisors.

Because gomphotheres lacked front teeth, they may have crushed the smaller fruits between molars and the bigger ones with their feet—just as zoo elephants today step on watermelons to access the pulp. Perhaps, too, ground sloths with widely spaced incisors that functioned like canine teeth punctured and ripped crescentia. The rhinolike toxodons, which originated in South America, would have been ideally equipped for the task. Possessing enormous incisors—"buck teeth" Paul Martin calls them—toxodons could have cracked even the biggest crescentia fruits. Because such fruits fall from the tree while still green and unpalatable, taking a month or so to ripen on the ground, the genus surely was not attempting to specialize on megafauna with trunks (gomphotheres), with long necks (liptoterns), or who easily balanced on hind legs (ground sloths). Horses and toxodons would not, therefore, have been beaten to the harvest.

We probably should leave out of our Pleistocene and modern scenarios most of the artiodactyls. Cattle, deer, and their ilk lack upper incisors. Instead, they graze and browse with lower incisors that cut leaves and stems against a horny plate on the upper jaw. More of a deterrent is the sweetness of crescentia pulp. Artiodactyls (including camels) rely on forestomach fermentation of plant fibers. Then as now, ruminants would have been repelled by the excessive sugars in crescentia pulp. Sugar-rich foods dangerously acidify a rumen, threatening the survival of cellulose-eating microbes who prefer an alkaline home and sometimes leading to ulcerous sores in rumen tissues. Sure enough, when Dan Janzen cracked some crescentia fruits and tested the open halves on cattle, the animals showed no interest in the contents.

Horses thus may not only be excellent surrogates for the missing Pleistocene partners of crescentia. They may be the sole surrogates. Because they swallow the pulp without much chewing, they do not injure the small seeds. Janzen reported that the same proportion of seeds that will germinate when taken directly from pulp (97 percent) will germinate when harvested from horse dung. He also observed that sapling crescentia trees are "commonplace" in ranch horse pastures and "extremely rare" where horses are absent. For all these reasons, the fruit of *Crescentia alata* does indeed qualify as a neotropical anachronism.

The pod of a legume tree, *Enterolobium cyclocarpum*, was the candidate anachronism subjected to the most rigorous testing. Janzen reported his results in a pair of papers published in the journal *Ecology* just a few months before the anachronism hypothesis appeared in *Science*.[13] He conducted the research during the summers of 1978 and 1979. Experimentation was simple but labor intensive. Janzen fed pods of the guanacaste tree (commonly known as monkey ears) to ranch horses under carefully controlled conditions. During the feeding sessions, he noted the number of seeds rejected rather than swallowed with the chewed pulp (paper number one). A second objective was to study how long it took the seeds to emerge, how many were killed en route, and how many of the survivors were scarred sufficiently to germinate promptly (paper number two).

Using ranch horses as proxies for the extinct megafaunal dispersers of native seeds may not, in hindsight, seem like a particularly novel experiment. At the time, however, it was. Janzen wrote in a 1981 paper, "There has long been a strong bias among field biologists against studying the two most easily studied species of large herbivores: ranch cattle and horses."[14] He continued, "It is indeed ironic that there is not a single paper in *Ecology, Journal of Animal Ecology,* or *Journal of Ecology* on the field ecology of the two large herbivorous mammals that have had the greatest impact on habitats the world over, and are our only chance even to begin to understand the impact that the New World Pleistocene megafauna must have had on New World vegetation."

Paul Martin encouraged Janzen in this way of thinking. Six months into their collaboration, Martin sent a handwritten letter to Costa Rica, where Janzen was just beginning his experiments with horses as seed dispersers. "An amazing lack of interest in the diet of Mexican cattle," he wrote, "is endemic among both animal science experts and 'pure' ecologists in this part of the world. I've come to appreciate the range cattle of Organ Pipe Cactus National Monument [southern Arizona], which all conservationists profess to hate, or at least insist be removed. I see cattle as a rich opportunity to learn about the interaction of desert plants and the bovid rumen."[15]

The most sophisticated equipment Janzen used in his horse experiments were canvas duffel bags reengineered to serve as dung catchers. Seeds were sorted by hand: "The dung was put into solution by running water into a bucket half full of dung. After lumps were thoroughly broken up by hand, the thin water-dung mixture was poured off and the *Enterolobium* seeds collected from the bottom of the bucket."

All seven horses used in the trials "readily ate hand-offered guanacaste fruits as familiar and desirable objects. All horses were likely to steal fruits from untended bags." Seed-rejection ranged from 38 percent, for the horse named Tinto, to Negrito's 75 percent. Janzen speculated that seed spitting would be reduced under natural conditions in which horses would tend to encounter fallen pods in herds. "Inter-horse competition" would encourage each horse to eat faster, rejecting fewer seeds in the process.

The results at the other end showed that it took a minimum of two days and a maximum of more than sixty for a seed to pass through a horse. Under natural conditions in which a herd might encounter a lone guanacaste tree and forage on its fruits intermittently for a day or two while also taking in grasses, the seeds would likely be defecated over a period of several weeks. This pattern of extending dispersal in space and time is to the plant's advantage. A substantial number of seeds were, however, killed en route during the experiment. Depending on the horse, 44 to 91 percent of the seeds failed to resist the digestive juices sufficiently to emerge in a viable state.

Does spitting out half the seeds and destroying more than half those swallowed make the horse a poor dispersal agent? Far from it. Consider the alternative: absent horses, guanacaste fruits rot beneath the parent tree. The point of producing thousands of seeds each fruiting season is to ensure that, sometime during the tree's life, at least one will germinate in a favorable spot and survive long enough to reproduce successfully in turn. A single guanacaste tree may live for more than two hundred years. If during those two hundred years, just one seed every few years survives long enough to produce offspring of its own, the grandparent tree should be judged a phenomenal success. By that standard, a disperser that swallows half the seeds and, of those, defecates a third in viable condition is a very fine partner indeed.

An African analog is instructive. The umbrella thorn, *Acacia tortilis*, is the picturesque, umbrella-shaped tree that frames sunsets on the wildlife-rich savannas of southern Africa. Its pods ripen during the dry season when other food sources are scarce. The pods and their pulp have a high nutrient content, and the seeds are protected by a tough coating. The pods of umbrella thorn do not rot beneath the parent. Giraffes pluck them from the spreading canopy, and lesser ungulates eagerly search the ground. Giraffes, kudu, and impala—all cud chewers—are considered effective dispersal agents, yet of the seeds initially swallowed, more than 90 percent are spit or crushed when these ruminants chew their cud.[16]

Although there are some instances of plant-animal partnerships (usually pollination rather than fruit dispersal) that entail a one-for-one match, evolution is usually "diffuse."[17] That's one reason why Janzen and Martin proposed a "megafaunal dispersal syndrome" rather than a gomphothere syndrome or an extinct horse syndrome or a ground sloth syndrome. Dispersal mutualisms aren't exclusive to a single species or even family of animal, though they may be exclusive (or nearly so) to much higher taxonomic levels, as in the bird dispersal syndrome or the ant dispersal syndrome. If the gomphotheres or ground sloths didn't arrive in time to harvest ripe guanacaste fruits still hanging on the tree, then the notoungulates or horses got them after they fell.

Seeds hoping to survive passage through a digestive tract must therefore accommodate a diversity of grinding mills and microbial fermentation vats. This creates complicated selection pressures. A plant must furnish its progeny with a casing tough enough to resist or deter the assault of a variety of dental batteries. The seed coat must be thick enough to retard germination for several to many days in a variety of moist and sometimes acidic environments—including rumens, stomachs, small intestines, cecums, and colons. Yet the protective coating must not be so thick that, once passed, the seed fails to germinate during the intended season of warmth or rain.

Given sufficient time and a degree of consistency among the multifarious selective forces, evolution should produce a fruit-seed combination that is recognizably adaptive from a gestalt and natural history perspective. Such judgments may be complicated by the tendency of plants to "satisfice" rather than optimize their traits.[18] Nevertheless, legume pods that rot on the ground are far from satisficed, but to Janzen's way of thinking, those that attract seed-spitting horses have found their intended match.

Native horses and other big mammals that once dispersed guanacaste seeds in their dung were ecological partners in yet another way. These megafauna prevented seed predators and parasites from devastating the crop. In this aspect of the guanacaste story, too, Dan Janzen broke new ground. In 1970 and 1971 he proposed that dispersal beyond the canopy of the parent is important not only to prevent intergenerational competition and to encourage colonization of new habitat, but also to help seeds escape the attentions of seed predators—be they rodents, parrots, or beetles—and to help seedlings escape detection by leaf-eating insects bred on the foliage of the parent.[19] In setting forth this "escape hypothesis," Janzen shows a predisposition that would gel into the anachronism idea six years later. Janzen wrote, "Very often the interaction that selected for the traits still displayed by animal and plant is now missing, making those traits seem maladaptive or neutral in significance."

By scattering dung over a wide area, fruit-eating mammals help seeds elude seed predators and parasites. Spread out, and sometimes buried deep within dung, the seeds are more difficult to find. In addition, some predators and parasites are deterred by the dung itself. Beetles of the family Bruchidea will deposit eggs not only on the surface of intact pods but also on any seeds they find exposed on the ground or on the surface of dung. They cannot, however, oviposit on seeds deeply embedded in dung.[20] Fruit-eating mammals control populations of parasites, too, by consuming fruits before the wormlike larvae have matured into beetles.[21] The larvae are killed by digestive juices that penetrate the seeds through the very holes the larvae had bored to gain access to seed nutrients.

Guanacaste seeds benefit from both the dispersal and the escape services offered by ranch horses. The triangular relationship among the guanacaste tree, ranch horses, and bruchid beetles has an African parallel. Again, it is the umbrella thorn tree that serves as analog. These acacia pods, as well as the pods of several other species of acacia trees, are avidly eaten by African ungulates, which not only scatter the seed but also kill bruchid beetle larvae inside.[22]

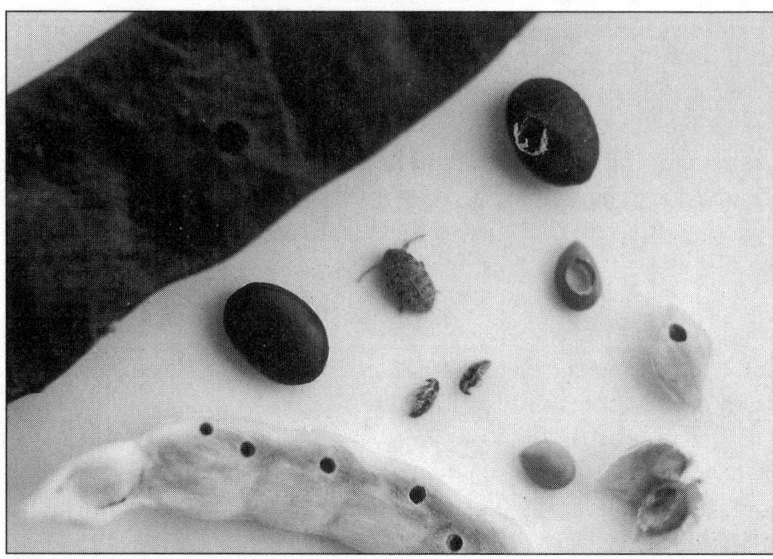

SEED PARASITES OF LEGUME PODS. Bruchid beetles of various species deposit their eggs on legume pods throughout the world. A hatchling bores into the pod and thence into a seed where it feeds. Before changing into an adult beetle, the larva bores a circular hole in the coat of its now-hollow seed to facilitate escape after metamorphosis. The newly emerged beetle drills another exit hole through the pod and begins the cycle anew. The pods of honey locust (top) and of mesquite (bottom) show exit holes. Exit holes are also evident in several seeds. The seed of mesquite is contained in a tough case (endocarp), whereas the honey locust seed lies bare in the pod. The two kinds of bruchid beetle shown here correspond, by size, to the two kinds of seed.

Legumes throughout the world are prone to attack by bruchid beetles. Six months before my visit to the Arizona herbarium, I received in New York a package of mesquite pods from Tucson. Paul Martin had collected them around his home in February, months after they had dropped to the ground. Because Martin lives in an urban district where there are no ranch horses or gomphotheres to call upon, the mesquite pods become a nuisance. A year's worth of pods from even a single tree can fill a dozen yard waste bags. The 1998 cleanup had already taken place by the time I asked Paul for samples, but enough remained to fulfill my request. All the pods I received in February were marred by at least one little hole, chewed away by an adult bruchid beetle exiting the pod. Some specimens had a hole over every seed. The following summer I collected freshly ripened pods of mesquite myself, as wild mesquite grows within a day's drive of our summer home in New Mexico. Placing the pods in jars, I waited for the beetles to emerge. The photograph on page 38 show the results.

You will notice honey locust in the picture too. Hanging on the tree after leaf fall has bared the branches, the fruit of honey locust looks strikingly like the pod that prompted Dan Janzen to begin thinking in a new way. Central America's *Cassia grandis* bears one of the longest fruits in the neotropics. Honey locust's pod can claim that distinction for temperate regions of North America. As a tribute to this remarkable tree of my homeland—and because its seed, conveniently, is one centimeter in length—a honey locust seed will serve as the scale bar in many of my close-up photographs of anachronisms to come.

How Honey Locust Got Her Honey

"So, what are you writing about now?" asked one of my friends from high school, whom I was visiting during a July trip to the suburbs of Detroit. I gazed out her living room window to clear my head for a long explanation—only to notice across the street a grand old honey locust loaded with young pods.

"Well, I'm working on a book about how honey locust—like that tree over there—got her honey."

The "just-so" style of storytelling used by Rudyard Kipling for his African fables (how the elephant got its trunk, how the leopard got its spots) is also a handy way to think about the evolutionary trajectories of anachronisms. How *did* the fruit of honey locust—and mesquite and guanacaste and *Cassia grandis*—come to possess sweet pulp surrounded by a tough pod? Just who was honey locust attempting to enlist as an emissary for her seeds?

A few words about vocabulary are in order before proceeding. The *her* in the preceding paragraph is scientifically more precise than the pronoun *its*. Honey locust is dioecious, producing male flowers and female flowers on separate trees; pods, of course, are produced only on female trees. Ergo, "her."

The phrase *trying to enlist* (as well as my earlier usage of *intends* in reference to a plant's dispersal "strategy") is simply an economical, if backhanded, way to refer to the forces of natural selection that shape the features of organisms who really aren't intending to do anything out of the ordinary. It is the lineage that produces variations through time, not the organism. And it is the environment (including other organisms in the environment) that determines which variations yield the most offspring that in turn successfully reproduce. This practice of using teleological language—the language of purpose—is widely, if grudgingly, accepted in evolutionary biology. We all know what we mean. It is helpful to ensure that our readers do too.

In a 1971 paper, for example, Dan Janzen compromised by using teleological language but embedding it in quotes. He wrote, "Seeds 'intended' for a dispersal agent may well be killed by another."[23] Later he explained his departure from scientific decorum: "May I also suggest that the reader not be dismayed by a terminology that might best be termed evolutionary shorthand. The statement, 'It is likely that heavy investment in chemical protection of their few leaves is worthwhile,' may be transliterated to read, 'Those mutant genotypes that produce toxic foliage without resources that are used for other processes in the original genotype have a higher relative fitness in plants with small leaf crops than in those with large ones.'"[24]

A prepublication draft of Janzen and Martin's 1982 anachronism paper included teleological language, but the paper was rewritten to eliminate these transgressions.[25] In the published version, for example, the authors refer to plant reproductive traits as having been "molded and maintained by complex interactions in which the gomphothere, with its huge stomach, massive molars, and peripatetic behavior, plays a central role."

Paul Martin flagrantly violated the etiquette of his guild when, in a 1992 essay about anachronisms in a popular magazine, he wrote: "Whose foot is devil's claw intended to intercept?"—which brings up another locution that may require justification: the matter of *who*. In this passage Martin uses *who* in a way that offends few scientists. Animals are enough like people to be awarded this implicit recognition of personhood. But *who* is not tolerated for plants; *which* or *that* is de rigueur. I shall nevertheless occasionally depart from the standard when referring to a plant. I can lean on Lynn Margulis for support; in her book *What Is Life?* she regularly uses *who* to refer not only to plants but (gasp!) bacteria. More to the point, it seems silly to replace *who* with *that* in a phrase attributing something as sentient as "intending" or "remembering" to plants.

But, then, plants do remember—though not by means of excitable neurons. The memory is retained in the genes. The genes of an avocado tree do not remember the journey through the intestinal tracts of generation upon generation of gomphothere and the rich environment in which generation upon generation of avocado seeds have germinated. The genes do not remember gomphotheres at all. But they do recollect that an oily, aromatic flesh surrounding a poisonous seed is the way their ancestors reproduced, and the way that they shall too.

There is one more reason I have elected sometimes to use so-called anthropomorphic language. The biodiversity crisis compels it. Anything scientists and science writers can do to promote a sense that other organisms are not lesser, dispensable beings is a virtue in these times. We can either elevate nonhuman organisms to our own exalted status or linguistically demote humans to equality with other organisms. Referring to honey locust as *she* is surely less offensive than the alternative. Jane Goodall (1998) writes that when she submitted her first paper for publication, the editor changed every *he* or *she* that referred to a chimpanzee into an *it*. Goodall stood her ground. She reflects now, "The paper, when finally published, did confer upon the chimpanzees the dignity of their appropriate genders and properly upgraded them from the status of mere things to essential beingness."

Back to our story: Janzen and Martin hypothesized in 1982 that the pod of honey locust, *Gleditsia triacanthos,* along with the fruits of several other North American trees, is anachronistic. This was just an aside in a paper centered on the fruits of the Costa Rican dry forests. That same year Janzen did, however, elaborate on the story of honey locust in an imaginative essay, "Famished Mammoths," which he wrote for the magazine *Garden*. "On a fine fall day in Kansas 10,000 years ago," he began, "a mammoth on its way to a creek stopped at the edge of a forest to sample newly fallen, slightly sweet honey locust fruits."[26]

Janzen's story is a playful overview of the subject, but it lacks detail. Fortunately, honey locust has attracted a good deal of scientific scrutiny because of its forage and ornamental value. We need no fresh experiments or observations to detect, in published reports, features of honey locust that strongly indicate anachronism.

One of these features is size. There is no point in building a fruit any larger than the gape of one's intended partner. If a potential disperser has to bite off a chunk, or if a section of fruit hangs outside its mouth, seeds may slip away or be eaten around. Because honey locust pods are commonly as long as my forearm, it is safe to infer that rabbits and deer did not exert strong selective pressures on the ancestral plants. No animals with mouths big enough to take in a pod whole are now native to the home range of honey locust.

Honey locust is exclusively a North American tree. A second species of the same genus *(Gleditsia aquatica)* is also found on this continent, but its small, pulpless pod is clearly not food for megafauna. Eleven other species of *Gleditsia* are native elsewhere: one in South America, another in northern Iran, a third in northeastern India, and the remainder in eastern Asia.[27] Ten of those eleven species bear pulpy pods like that of honey locust. For

HONEY LOCUST AND HER NEW PARTNER IN EVOLUTION. Strung along the top are a usual number of seeds in a honey locust pod, such as the one resting on my arm. The tree has doubly pinnate leaves typical of the legume family.

North America's *Gleditsia triacanthos*, the native range—that is, where the plant is thought to have grown before colonizing Europeans and their livestock began rearranging the vegetation—is the central United States.[28] Bounded on the west by the treeless Great Plains and on the east by the dense forests of the Appalachians, honey locust inhabited prairie openings in midcontinental woodlands as far north as Michigan and south into Texas.

Because many North American plants adapted for temperate climates were pushed far to the south during the several pulses of glacial advance that marked the Pleistocene, one should not expect that an anachronistic tree of the temperate zone must necessarily occupy a range today more constricted than it occupied, say, eighteen thousand years ago (the last glacial maximum). Some species of trees scatterhoarded by squirrels or birds are still rebounding from the Ice Age, extending their ranges. Rather, the geographic indicators of anachronism would be patchiness and ease of naturalization. Honey locust is "a minor component in riverine forest associations in the eastern United States."[29] Yet it has been successfully planted in all forty-eight states.[30] More, it has become naturalized—that is, escaped from cultivation into the wild—in many places, including lands east of the Appalachians.

That wild honey locust is usually found in floodplains is another clue. If the plant has lost its ability to dependably move seeds uphill, then the usual direction for them to go is down. Indehiscent legume pods, including honey locust, are excellent floaters—not because flotation was the plant's intent but because sealing a pod to prevent water loss and to keep out all but the most insistent seed predators and parasites has the ancillary effect of keeping out water. Most wild honey locusts are found in floodplains, but the tree is known to grow in uplands too.[31] Seedlings establish under a range of temperatures and all but the driest conditions.[32]

These traits enabled honey locust to claim as habitat urban areas throughout the United States during the past half century. Douglas Still, coordinator of street tree planting for the Manhattan Borough of New York City, told me that honey locust is "a good performance tree." Among its attributes is an ability to grow in both wet and dry soils. The tree also tolerates a range of soil acidities, including the high alkalinities produced by the calcium in concrete. The tree can handle road salt in winter, high heat in summer, and intense soil compaction (hooves perhaps?) year-round. For all these reasons, honey locust thrives along the streets of New York.

Honey locust does have a problem. One cannot harvest pods in the wild and expect the planted seeds to germinate any time soon. This feature, too, is an indicator of anachronism. Before honey locust seeds will germinate, they must be scarified—that is, the tough seed coat must be breached sufficiently to allow water to penetrate and swell the seed. "To obtain successful germination, seeds must be scarified and forced to break seed coat dormancy. This can be accomplished via immersion in concentrated sulfuric acid for one to two hours, hot water (82°C), or by mechanical means."[33] Another propagator of honey locust advises, "The seed requires damage before it will germinate and is believed to be effectively scarified by passing through the digestive systems of cattle. Scarified seed has ten times the emergence of unscarified seed. Laboratory experiments showed 100 percent of scarified seed germinated within sixty days. Unscarified seed field sown in 1985 emerged in 1987 and even into 1988, suggesting a dormancy period of up to three years under field conditions."[34]

Here is another report: "When seeds were collected from the pods found on the ground in the winter or next spring and planted in the greenhouse, only very rarely would a single seed germinate even after a year or longer. . . . By filing a line through the seed coat until I could see the cotyledons, I found I could get almost 100 percent germination. . . . In nature, probably various agents—for example, freezing and thawing, or, passage through an animal's gut—may enhance germination."[35]

One of the nice things about this ghosts project is that it has offered a chance to engage in a little kitchen-sink experimentation. Nothing rigorous and statistically significant, mind you, but a level of testing and observation by which I can experience phenomena described by others. In mid-February 1999, I immersed in a cup of tapwater three honey locust seeds removed from pods that I had scavenged from the sidewalks near my New York apartment. I kept those seeds wet, periodically changing the water. Three months passed, and the immersed seeds were still very hard and seemingly unchanged. It was mid-May, and I could wait no longer; in a week I would be heading for a summer in New Mexico. So I scarified one of the seeds, leaving the other two as controls. I was struck, first, by how difficult it is to create a scar, even after a seed has had three months of soaking. These seeds are nothing like the domesticated pinto, black, and kidney beans that soften overnight in a bowl of water. Back and forth I sawed across the honey locust seed with a serrated kitchen knife. I stopped sawing only when a line of beige appeared through the dark brown seed coat. I then returned the seed to its cup of water. The very next day, the scarified seed had swollen to twice the size of its companions. By the third day, a robust root had appeared.

Yes, indeed. The seed coat does retard germination to a degree that appears maladaptive in today's environment. Might that hardy seed coat have evolved to withstand mastication and digestion?

The toughness of the seed coat and the bigness of the pod are two traits of honey locust that seem to be overbuilt for current dispersal opportunities. There is a third trait: the honey. Why waste energy manufacturing a sweet, protein-rich pulp unless that pulp is meant to attract an animal?

The sugar content of the pods of wild honey locust can be as high as 35 percent.[36] One advocate of honey locust agriculture calculated that a field of mature honey locust trees would generate in their pods as much harvestable sugar as could be extracted from the same size field of cultivated sugarcane.[37] Sugar is not the only attraction. The same advocate reported that a 1942 study by the Alabama Agricultural Experiment Station determined that honey locust pods grown on a single acre were equivalent in overall nutrition to 105 bushels of oats. More recent work indicates that the nutrient value of honey locust pods approaches that of whole alfalfa plants.[38] Consider, too, the season of availability of these pods: late fall and winter, when the grasses many herbivores depend on have withered and the leaves that sustained big browsers are no more. A single honey locust tree will drop its pods over several months. Many linger on the branch, accessible to anyone who can reach them, even when the snow is deep.

"Each day a few of this tree's fruits fall to the ground, and each day the tree is visited by the members of the local horse herd, three bison that know the tree's exact location, and one camel. An occasional mastodont or mammoth also stops and snuffles around the tree." This passage comes from Dan Janzen's "Famished Mammoths" essay. I would have had the mammoth do more than snuffle, however. African elephants are known to harvest acacia pods not only by extending their trunks but also by rearing up on hind legs and vigorously shaking down the bounty.[39] Even a horse might stretch its reach beyond the customary. "Our old buggy horse used to stand on his hind legs to pull the [honey locust] beans off," reads one account.[40] Honey locust pods must have held great appeal for a wild ungulate in winter, long after the range grasses had sucked nutrients back into their roots (that's why hay is cut green and left out to dry) and especially after snowfall had blanketed the land.

Livestock favor the pods enough to have secured honey locust a visa to foreign lands. Sometimes, however, the receiving country gets more than it bargained for. *Gleditsia triacanthos* and some American species of mesquite *(Prosopsis)* have invaded so much of Australia's rangelands that these legumes are now considered pests. An eradication program for honey locust was initiated in 1993 because the tree "has the potential to smother pastures and replace native riparian vegetation."[41] Infestations of honey locust along riverbanks in Queensland owe to transport by floods. Those that spread beyond the floodplains must have hitched a ride in the rumen or gut of cattle.

Honey locust has also been introduced in Europe and New Zealand, apparently without unwelcome intrusion into wild landscapes—yet. In South Africa and Chile, however, this hardy tree has taken the same path as in Australia. Despite this cautionary experience, there is talk of introducing honey locust as a fodder tree into yet another realm: the foothills of the Himalayas.[42]

Janzen continues his story of honey locust, following the demise of its Pleistocene partners: "Because seed dispersal is minimal and seed predation is heavy, the honey locust now grows only in those habitats where a seed has a very high chance of becoming an adult plant. The honey locust has become a rare tree. The honey locust remains rare for a few thousand years—not enough time for it to undergo the evolutionary changes that could adapt its fruits to a new set of dispersers and its vegetative traits to the restricted habitats in which it finds itself."

Then the Europeans arrive. "The Europeans bring their horses and cows. These animals find in the fallen honey locust fruits the same kinds of dry legume fruits that their ancestors ate in the Old World tropics and subtropics; the horse in particular finds the fruits that its earliest ancestors grew up on." Honey locust "is soon found again over a wide geographic area and in high densities in a wide variety of habitats."

We are so unaccustomed to looking into the past for explanations of the present that scientific treatment of honey locust dispersal has, except for Janzen's account, ignored the Pleistocene megafauna. If not megafauna, then who?

The wind: "The purple pods cling and rattle in the wind long after the yellow leaves have fallen. One by one, they are torn off, their S-curves tempting every vagrant breeze to give them a lift."[43] Another commentator writes that the pods "are artfully twisted so they roll when tossed off the tree in seed dissemination."[44] It appears that this charming story comes from one of the great botanists of the nineteenth century, Charles Sprague Sargent. Later authors simply took him at his word. Here is how Sargent described, in 1890, honey locust's mode of seed dispersal: "The pods contract in drying with a number of cork-screw twists and without this provision they would remain where they fall under the trees, but the pods thus twisted roll like wheels. Being very light they are blown for great distances over the frozen ground and especially over the snow. The obstacles they are obliged to overcome in their journeys probably help to break open the pods and liberate the seeds."[45]

The biggest extant native herbivore in eastern woodlands has also been pegged as the partner of honey locust. "Animals are probably the chief dispersal agents, and animals, particularly cattle, are known to eat the pods. Cattle, of course, could not have been responsible for dispersal until fairly recently, so perhaps deer were once the chief agents of dispersal."[46] Deer, however, are inclined to spit rather than swallow seeds, and the fecal pellet of a deer is little bigger than the seed in question.[47] There were no representatives of the deer family (Cervidae) in the Western Hemisphere until about five million years ago. The post-Pleistocene fauna of the midcontinent yields no plausible dispersers.

During the 1990s Andrew Schnabel wrote more technical papers on honey locust than anyone else. He knows too much about honey locust to accept wind or deer as effective dispersers. In fact, he reported that, except where cattle are present, very little seed is dispersed. During a field study conducted in Kansas, Schnabel and coauthors found that "few seeds were dispersed greater than fifty meters from maternal trees."[48] The primary dispersal agent, in his view, is gravity, but "secondary dispersers include horses, cattle, and deer, as well as a variety of small mammals. It is common in this population [in Kansas] for dispersers to sit directly at the base

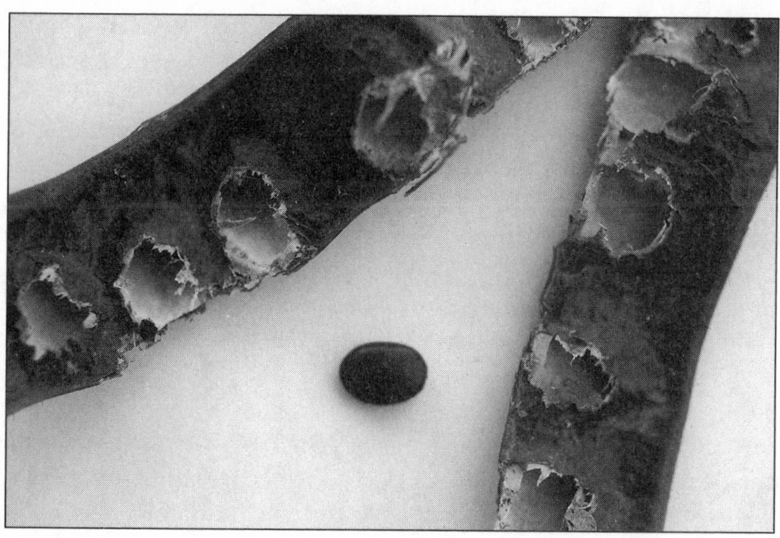

The work of a seed predator. These honey locust pods lost their seeds to a mouse.

of trees while extracting seeds from fruits."[49] (From an anachronism standpoint, any animal that wastes good pulp while extracting just the seeds is a seed predator, not a disperser.)

I asked Schnabel whether he was familiar with the anachronism idea. To my surprise, he was. He encountered the Janzen and Martin paper while still a graduate student at the University of Kansas. "The issue came up not with *Gleditsia*," he recalls, "which has some obvious present-day seed dispersers (e.g., deer), but with *Maclura pomifera* [osage orange]." He continues, "No, I don't think that honey locust is limited evolutionarily by low seed dispersal. That is, there seems to be considerable anecdotal evidence that honey locust seeds are dispersed over long distances (one kilometer or more), albeit rarely. As for *Maclura pomifera,* I don't know of any native species that disperses seeds over an appreciable distance. My casual observations are that *M. pomifera* does not spread into pastures and old fields as fast as *G. triacanthos*."[50]

Schnabel's response was not unexpected. My experience in reading papers and talking with biologists and ecologists is that anachronism is all too often interpreted as a binary. Either a fruit is an anachronism or it is not. Either a fruit is dispersed or it is not. If it is dispersed, even feebly, then it cannot be an anachronism. End of discussion.

Following Janzen and Martin, I believe that an observation of feeble dissemination of a fruit type marks the beginning, not the end, of study. This is precisely when the questions become interesting. Very few plants that merit our attention in this book will be as underdispersed as osage orange. The rest will profit from the occasional seed-spitting deer or frightened rabbit who runs off with fruit in mouth. Nevertheless, something about those associations strike us as far from economical. The fruit will appear overbuilt and underutilized. We will wonder why nature would have evolved something so inept. Why would natural selection dedicate so much resource for so little gain? The anachronism concept treated as a continuum is a focus of Chapter 4. There we shall see that most of the fruits identified by Janzen and Martin are anachronistic *to a degree*. Some are moderately so, others substantially, and a very few are anachronistic in the extreme.

There is one more facet to the story of how honey locust got her honey. We must turn to the question of origin. Fruit characteristics tend to be highly conserved through evolutionary time at the family and even the genus level.[51] This means that the fundamental features of a plant's dispersal strategy were shaped tens of millions of years ago when the families and genera of flowering plants were taking their form. Honey locust did not evolve *for* the Pleistocene megafauna, although the Ice Age mammals did maintain the legacy of selection for big, pulpy, and woody pods. If those mammals were not responsible for the origin of pod traits attractive to large mammals, who was?

Thanks to a "cladistic biogeography" analysis undertaken for genus *Gleditsia,* Andrew Schnabel has provided the information I need to begin to answer this question. In a 1998 publication, he and Jonathan F. Wendel reported that the oldest reliable fossil of the genus *Gleditsia* occurs in Oligocene sediments in North America that are twenty-three to thirty-five million years old.[52] *Gleditsia,* they surmised, was probably widespread in both Asia and North America by ten million years ago. What sorts of pod-eating megafauna were browsing or grazing North America then, and who would have shaped the fruit initially?

Horses and other equids were both speciose and abundant during the Miocene of ten million years ago. North America is, in fact, thought to be where family Equidae originated and diversified. Emigrants later spread to Asia, Europe, Africa, and South America. Today, only Africa is home to wild equids (zebras). North America's early equids, however, were no bigger than a small dog. The lineage may later have acquired a mutualism with honey locust, but we must look to other early beasts for selective shaping of the honey locust pod.

Brontotheres may have had an important early influence. Like horses, these herbivores were odd-toed hoofed mammals. And like horses, the earliest fossils of brontotheres hail from North America. The brontothere clan originated on this continent at least fifty million years ago, during the Eocene, and thrived for fifteen million years. Then they went extinct. At the height of their glory, however, some brontotheres were elephantine in size. So were the rhinos at their evolutionary peak. They, too, seem to have originated in North America about fifty million years ago and were common on this continent until just five million years ago. The five species hanging on in Asia and Africa today are mere remnants of a once mighty and ancient lineage. Many extinct rhinos looked like the rhinos of today, but some resembled hippos and others horses. The biggest land mammal of all time was a rhino. *Indricotherium* lived in Asia thirty million years ago, and it stood taller than a mammoth.

Bulk is not a prerequisite for height. Fossil evidence suggests that the camel family arose in North America thirty-three million years ago. *Aepycamelus,* which lived in North America thirteen million years ago, could have harvested a lot of honey locust pods before other animals got a chance. A greatly elongated neck and long limbs gave *Aepycamelus* the appearance of a giraffe. Might this camel once have been a sustaining partner for honey locust?

The elephant lineage did not originate in North America, but by fifteen million years ago ancestral mastodons were here. Then too, let's not overlook the chalicotheres, who lived fifty-five to twenty-three million years ago, and the protoceratids, who vanished just five million years ago. It would be fruitless to attempt a discriminating analysis of pod-eating potentials among these candidate dispersers of bygone times. Fossil evidence does not reveal details of digestive anatomy, physiology, or behavior. We might hope to find a seed in a coprolite, but whose turd is that anyway?

Nevertheless, we can draw some tentative conclusions, based on dentition and one crucial distinction in gut design that can be traced back millions of years. These matters become important in Chapter 8. Equally important, and far easier, we can distinguish fruits evolved to appeal to large mammals from those that forged mutualisms with bats or birds or monkeys or ants—or that chose simply to hitch a ride on the wind. What is it about honey locust that should have made this fruit attractive to megafauna now extinct in her homeland? Which traits work against successful recruitment of lesser creatures, like deer and rabbits? Paul Martin posed these questions in the late 1970s. As you will see, an understanding of animal-fruit interactions on the two continents where megafauna can still be found was crucial for his project. In becoming acquainted with the megafaunal fruits partnered with elephants and rhinos in Asia and Africa, we learn to spot anachronisms in our own backyards.

AN ULTIMATE ELEPHANT FRUIT FROM SOUTHEAST ASIA. Durian fruit is renowned for the fine flavor—but abominable odor—of its creamy pulp. The large seeds are encased in a matrix of yellow custard that is exceptionally rich. The husk is cloaked with spikes the size of pencil tips that deter pulp thieves and seed predators alike.

3

The Megafaunal Dispersal Syndrome

SUMMER EVENINGS, TYLER AND I LIKE TO SIT ON THE PORCH STEPS of our trailer in New Mexico spitting seeds. As the Summer Triangle appears over the canyon wall, we're lazily spitting cherry pits or watermelon seeds, sometimes dropping popcorn. By morning the area has been tidied up. The nocturnal litter squad provided by order Rodentia has taken care of the problem.

Any watermelon seeds or popcorn the rodents may have missed will be discovered by harvester ants. An imposing anthill rises behind our trailer, exactly (we are told by the previous owner) where it has been for at least thirty years. I sometimes notice an ant struggling with a kernel of popcorn, even a whole watermelon seed, as it crosses the cement stoop before commencing the final twelve meters of far more difficult terrain. If the tidbit came from our compost pile, a twenty-meter trek is required.

Seeds are being moved alright, but not in the ways and by whom the plants intended. Watermelon plants did not evolve plump melons in order for bipedal primates (or any primate, for that matter) to plunder the pulp, nor for rodents or ants to walk away with the seeds primates reject. Watermelon is of Old World origin; the fruit ancestral to our own overblown cultivars was intended for big beasts with big mouths that could crush the rind and swallow the mash, seeds and all. When eating watermelon, I am playing the role of pulp thief. When rodents or ants carry off seeds, they are seed predators. My spitting and their hoarding may occasionally result in a new watermelon plant, but all of us—ants, rodents, and humans—are backup dispersal agents at best. We are not the target mutualists. The fruit's attractions do not owe to the attentiveness of ant, rodent, or primate ancestors.

Target dispersal agents of fleshy fruits are those animals whom the color and aroma of the fruit's skin or rind, the taste and nutrition of its pulp, the style of presentation, and the schedule of fruit fall were evolved to attract. Target dispersal agents may crunch or digest some or even many seeds, but extracting nutrients from the seeds themselves (as in seed predation) is not their intent. Rind, pulp, and seed traits are evolved to attract these preferred dispersal agents while resisting pulp thieves and seed predators or parasites.

Field observations of who eats what, along with tests of seed viability in dung, can help identify target dispersal agents for one plant species or another. But a sign of a mature science—a science that has moved beyond cataloguing observations—is the establishment of broad principles. In this sense, the 1970s marked the advent of the science of seed dispersal ecology. Until then, dispersal modes of particular plant species had been noted as natural history, but little effort was made to develop general principles. To add to the confusion, before this time (and, alas, even today) ecologists failed to adequately distinguish target dispersers from backup dispersers, including pulp thieves and seed predators.[1]

"Give us a couple of hundred more years of careful close observation," Dan Janzen wrote in 1986. "We are at the stage parallel to medicine when it didn't know what red blood cells were or did, and still puzzled over the seat of the soul."[2]

Dispersal Syndromes and Their Animal Guilds

Dispersal syndromes—complexes of fruit traits that enable plants to disperse seeds—are ways to make science out of natural history. The least controversial syndromes pertaining to animal mutualists (as distinct from wind, water, or ballistic dissemination strategies) may be the *ant dispersal syndrome,* the *flighted bird dispersal syndrome,* and the *scatterhoarding syndrome.* To be recognized as a distinct syndrome, a complex must have an identifiable set of fruit traits and the kinds of animals attracted to the fruit must belong to a particular taxonomic group (as with ants or birds) or serve a particular ecological function (as with squirrels and jays that hoard seeds). Identifiable *guilds* of animals are therefore attracted to fruits of the corresponding dispersal syndrome.[3]

The key distinguishing feature of fruits for whom ants are the target disperser (and usually the only one) is possession of a tiny packet of lipid-rich food called an elaiosome, which is attached externally to a hard-coated seed. The food packet, not the seed, is the bait. The intact seed is hauled to the nest by workers; later, the elaiosome is eaten but the seed left unharmed. Sans elaiosome, the seed becomes trash, discarded in a midden nearby.[4] Melon seeds and popcorn may be plundered by large harvester ants, but such seeds were not designed *for* ants. In contrast, tiny seeds embellished with elaiosomes are wooing ants, just ants. Mice may make off with the goods, but that is not the plant's intent.

Many small-seeded woodland plants rely on ant dispersal. Wild ginger, *Asarum canadense,* is an ant-dispersed herbaceous plant of eastern North America. Arid lands susceptible to frequent fires (much of Australia and the fynbos region of southern Africa, for example) are also rich in plants that express the ant dispersal syndrome. Storage underground may be the only way some seeds survive fires.

The kinds of fruits that birds (not including flightless birds) are attracted to are usually small, with only a thin protective skin, and the colors are red or dark shades of blue or purple (sometimes white, though, or a patterned blend of these colors). Charles H. Janson contrasted this bird dispersal syndrome with a general *mammal dispersal syndrome.* Fruits categorized as mammal syndrome are bigger than bird fruits. No bird fruit can possess a tough rind or husk, but a mammal fruit might. Many mammals (especially those active at night) have poorly developed color vision, but unlike most birds, their olfactory senses are excellent. A mammal fruit may thus emit a strong odor when ripe but retain a dull coloration—brown or burnished yellow or orange—or even remain green.[5]

Because some bird-dispersed fruits are also spread effectively by mammals (huckleberries by bears), some ecologists prefer to make no distinction between fruits that attract birds and those that attract mammals, lumping all into a *vertebrate dispersal syndrome.* Others posit a continuum, with some plants exclusively attracting bird or mammal dispersers and others attracting both.[6] In a 1992 analysis of fruits in Spain borne by 910 species, Pedro Jordano found that the only trait that consistently distinguishes a mammal from a bird fruit is diameter: the bigger the fruit, the more mammals it attracts as preferred dispersers and the fewer birds. Unfortunately, Jordano did not consider color or aroma because they are not easily quantified. Quantifiable or no, color and aroma may offer the best way to distinguish mammal from bird fruits.

As a subset of the mammal dispersal syndrome, Carlos Herrera proposes a *carnivore dispersal syndrome,* which attracts the services of bears, foxes, raccoons, civets, and the like.[7] Fruits attractive to these carnivores, Herrera says, are large, many-seeded, pulp-rich, brown, and scented. They tend to fall to the ground upon ripening rather than remaining on the branch, and have pulp high in fiber and low in protein and minerals. Herrera concludes that carnivores are "legitimate" dispersal agents for such fruits. He notes, however, the problem of expecting current dispersal data to reveal much about the past. Did carnivores shape the fruit traits of these lineages tens of millions of years ago? Are carnivores thus the target mutualists? Would carnivores have had much access to ripe fruits in prehuman times, when large numbers of diverse megafaunal ungulates would have roamed the continents? Herrera speculates, "Carnivore dispersal represents the vanishing remnants of a more complex pre-Holocene plant-mammal seed dispersal system." He cites Janzen and Martin 1982 as support.

Herrera has noted elsewhere how fossils reveal that "the basic morphological characteristics of the fruits of modern species were present at their initial appearance."[8] Like the pod of honey locust, the fruits and seeds of many species and genera seem to have changed little since their plant families first appear in the fossil record as far back as the Oligocene (some thirty million years ago). In contrast, great have been the changes in the fox family, Canidae, and the bear family, Ursidae—neither of which had even begun to take shape in the Oligocene. Other carnivorous mammals, now extinct, occupied the ecological niches held by canids and ursids today.

Whether attractions for carnivorous mammals, frugivorous birds, or extinct large herbivores, vertebrate dispersal syndromes are today exceptionally important adaptations for plant reproduction. About a third of the fruits produced by woody plants in temperate deciduous forests rely on bird, mammal, and to a lesser extent reptile dispersal. In temperate coniferous forests, savanna woodlands, Mediterranean scrubland, and neotropical dry forests, just under half of woody species bear fruits targeted for vertebrates. The figure reaches seventy percent in subtropical humid forests of both Old and New Worlds, and ninety percent in tropical rainforests.[9]

Nut trees like oaks and beeches, whose fruits consist of a seed protected by an indigestible hull, cater to seed predators. Scatterhoarding squirrels and jays occasionally fail to claim all the nuts they stash for the winter. A bat dispersal syndrome is sometimes identified too (especially for figs), but monkeys take a lot of so-called bat fruits. A primate dispersal syndrome suffers the complication that some frugivorous primates (spider monkeys, woolly monkeys, tamarins) are seed swallowers, while others (vervet monkeys, colobine monkeys, yellow baboons) tend to be seed predators.[10]

Dispersal syndrome was an important part of the paper that launched anachronism theory in 1982. For a decade, Dan Janzen had been studying various fruit species in Costa Rica; the anachronism insight allowed him to reinterpret and thus satisfactorily explain heretofore puzzling fruit traits. Paul Martin unified these disparate fruits and their anachronistic

Strategies of the megafaunal dispersal syndrome. Domesticated varieties of three tropical fruits native to the New World demonstrate a range of pulp attractions and seed defenses. Ripe papaya fruit, *Carica papaya*, is soft enough to mash rather than chew, so the tiny seeds require no physical protection. A mammal that inadvertently crushes a papaya seed is, however, deterred by a sharp, peppery flavor—and thus the toxins so signaled. Primate fruit thieves (like us) can eat around and discard the concentration of seeds, thus foiling the papaya's intent. Avocado, *Persea americana,* produces a slippery and dense seed, whose potent toxins taste bitter to mammals. In contrast, the seeds of the canistel tree *(Pouteria campechiana)* have a mild flavor and are protected instead by a tough coating. All but the biggest frugivores could be expected to eat around or spit the seeds of avocado and canistel.

features within a new ecological category, which he called the *megafaunal dispersal syndrome.*

Dispersal syndromes of any flavor did not, however, sit well with Janzen.[11] In the early 1970s he cultivated an alternative way to think about such plant-animal interactions: "seed shadows."[12] His goal was to understand the dispersal contributions of all vectors. Their combined activities determine a plant's seed shadow—the configuration and density of final destinations of the seeds. Janzen thus rebelled against the then near-exclusive focus on primary dispersal agents, also known as "dominant," "legitimate," "efficient," or "effective" dispersers. For this book's evolutionary focus, however, the identification of primary dispersal agents—though now extinct—is crucial to the anachronism quest. Here the adjective *target* or *intended* is preferable. For anachronistic fruits, the target dispersal agents are not the animals interacting with the plants today; they are the primary mutualists from the past. So powerful a role did the megafauna play over evolutionary time in the reproductive success of the avocado and honey locust lineages, for example, that these fruits appear as if they were designed *for* the great beasts.

Janzen remains opposed to categorizing particular fruits under this or that syndrome. Rather, he prefers to investigate the entire coterie of big and little beasts and nonbiological processes that scatter seeds greater or lesser distances from a parent tree and with more or less success. One can determine only retrospectively which parts of a seed shadow are most conducive to reproductive success. Which sites, for example, are least susceptible to discovery by insect, rodent, and avian seed predators? Which sites offer conditions adequate for germination and continued growth of the seedling and sapling?

Seed shadow and dispersal syndrome both have a presence in the 1982 paper. Janzen contributed the former, Martin the latter. "Extinction of the Pleistocene megafauna would eliminate some of a tree's disperser coterie and thereby excise part of the tree's seed shadow," Janzen wrote, while agreeing to set aside his preferences and go along with Martin's identification of a megafaunal dispersal syndrome. "Paul is more organized than I am," Janzen recalls. "I cannot impose my own brand of phenomenon dissection on others." Martin's portrayal of the anachronism idea in the context of a new syndrome has, in fact, "helped people to think."[13] Indeed, the megafaunal dispersal syndrome has proved so useful that African ecologists have sometimes cited the paper not for the anachronism idea but for the syndrome.[14]

Paul Martin began using the term *megafauna* in his Pleistocene overkill work of the 1960s. He defined megafauna as animals that weigh (or weighed) more than a hundred pounds, having chosen this threshold because "it is easy to remember and roughly the weight of a small adult female person."[15] Pounds are no longer acceptable in science, so we are faced with a number not so easy to remember: forty-five kilograms. A rule of thumb helpful in North America is that *megafauna* means any animal at least as big as a pronghorn or deer.

Since then, Norman Owen-Smith has used the term *megaherbivore* (distinct from *megafauna*) to refer exclusively to the biggest of the big—those animals (all herbivores) weighing more than a thousand kilograms.[16] The only terrestrial megaherbivores alive today are in the Eastern Hemisphere: elephants, rhinos, and hippopotamuses. Giraffes and North American bison marginally qualify, as some males may exceed a thousand kilograms. As it turns out, all but one of the traits proposed by Janzen and Martin as signaling the megafaunal dispersal syndrome would hold whether the coterie of dispersers consists of mastodons and giant ground sloths or of mammals no bigger than deer. That distinguishing trait is fruit size (and, to a lesser extent, seed size and presence of a tough rind). As we shall see, some of the most compelling examples of candidate anachronisms could have been taken whole into the mouths of only the largest extinct beasts.

Traits of Megafaunal Fruits

In their 1982 paper, Janzen and Martin identified a dozen fruit traits indicative of the megafaunal dispersal syndrome, as displayed in whole or part by the fruits of thirty-seven species of trees or large shrubs native to the lowland deciduous forest of Costa Rica. Wherever in the world this syndrome occurs in plants yet megafauna are missing, one should investigate whether the fruits are anachronistic. What combinations of fruit traits indicate the megafaunal dispersal syndrome?

The most important traits fall into two broad categories: *mammal traits,* which distinguish mammal dispersal from other classes of animal dispersers, and *megafaunal traits,* which distinguish disperser size. In addition, *ecological indicators* show whether a fruit possessing the traits of the megafaunal dispersal syndrome suffers dispersal problems. These traits and indicators should be treated as cumulative for discerning anachronisms.[17]

CATEGORY I: MAMMAL TRAITS. *A fruit attractive to mammals, especially to the many who are color-blind:*

1. **Ripens to a nondescript color.** Fruits that are green when still growing and inedible may stay green when ripe or else change to brown or dull yellow and orange hues.
2. **Has a penetrating aroma.** Especially fruits that remain green when ripe will develop a strong odor to signal ripeness and to help mammals locate them.
3. **May consist of fibrous pulp.** Birds lack microbial fermentation vats, so fibrous flesh (like a pomegranate from the Mediterranean or breadfruit from Asia) provides little nourishment for the weight that would have to be taken on. Herbivorous mammals, in contrast, extract food value from fibrous plant matter.
4. **May protect the pulp in a tough rind or pod.** Birds are deterred by the rinds of oranges (native to Asia) and passion fruit (native to Brazil), as well as the stiff pods of all Old and New World indehiscent legumes.

Mammals, especially if large, are undeterred by these protective devices and may even be able to extract food value from tough components.

CATEGORY II: MEGAFAUNAL TRAITS. *A fruit attractive and accessible to large mammals but that deters or is overbuilt for small mammals:*

5. **Is large.** Size is the key feature that distinguishes megafaunal fruits from fruits of the general mammal syndrome. They have evolved to be consumed, without seed loss, by animals with a gape large enough to take such fruit into the mouth whole rather than piece by piece.[18] Plants that produce very large fruits may adopt the habit of "cauliflory," bearing fruits on or close to the trunk and stout branches. Cauliflorious fruits (breadfruit in Asia; papaya and *Crescentia* in the neotropics) are thus especially indicative of the megafaunal syndrome.
6. **Is indehiscent.** Large fruits must be indehiscent, meaning they retain their seeds upon ripening. Pods or capsules, no matter how big, that spill, eject, or dangle seeds are hoping to attract something other than a big mouth.
7. **Possesses seeds that deter or elude dental grinding.** The seeds, if large, may be protected by a thick, tough, or hard endocarp (pit) or seed coat that withstands the grinding of molars—or, more usually, dissuades the animal from even attempting to crush them. If the fruit is a woody pod, the seeds should be especially tough, because such fruits require sustained mastication. Another, less common way to protect vulnerable seeds is to provision the pulp with woody obstacles that frustrate mastication (e.g., *Guazuma ulmifolia*). If the seeds are large but not physically protected, they will be chemically protected by bitter, peppery, or nauseating toxins. Big mammals quickly learn not to chew such seeds. This chemical strategy is most common in fruits whose pulp is so soft that chewing is unnecessary (e.g., avocado and papaya). Whether physically or chemically protected, seeds of megafaunal fruits may be difficult to separate from the pulp (e.g., mango and persimmon) to deter seed spitting. If very small (e.g., figs and cacti), seeds may easily elude the molar mill of large animals.
8. **Possesses seeds that may benefit from (even require) physical or chemical abrasion in order to germinate.** If a lineage adapts to attract megafaunal dispersers, it may need to envelop its seed in a protective coating or pit strong enough to survive contact with teeth and digestive juices. Absent such abrasion, the seeds of stiff-podded legumes (e.g., honey locust and Kentucky coffee tree) may fail to germinate or may be seriously delayed.

9. **May fall to the ground when ripe or nearly ripe, especially for upper canopy trees of tropical environments.** Any tree of the upper canopy that produces fruits intended for megafauna must drop the fruits when ripe. Even elephants cannot reach to the tops of tall trees. Small trees, however, have a choice. In both the Old and New World tropics, monkeys can be formidable pulp thieves or seed predators. In tropical forests, therefore, the best strategy may be for fruits to drop upon or just before ripening. In temperate climates where monkeys are absent, a better way to avoid losses to pulp thieves and predators may be for fruits to remain on the branch. Here ground-based rodents and other ineffectual dispersers pose a greater threat than do arboreal thieves and predators. As we saw, honey locust of temperate North America exemplifies this approach; its ripe pods may remain on the tree for several months.
10. **Looks, feels, smells, and tastes like fruits known to be dispersed primarily or only by megafauna, where megafauna still exist.** This final megafaunal trait sums up the rest. In combination with ecological indicators, a comparative test may be one of the best ways to discover and screen candidate anachronisms.

CATEGORY III: ECOLOGICAL INDICATORS. *Signs of missing dispersal partners include:*

11. **The fruit either rots where it falls or is ineffectually disseminated by (or appears overbuilt for) current dispersal agents.** If the fruit rots beneath the parent tree, or if pulp thieves or seed predators are gleaning the bulk of the harvest, then missing partners should be suspected.
12. **The plant is more common where horses or cattle are present.** If the plant in question is rare where livestock are lacking and more abundant where horses or cattle have been introduced, a reasonable conclusion is that the fruit possesses anachronistic traits. If fallen fruits are avidly eaten by introduced horses or cattle, and if at least some seeds are swallowed whole and are still viable when defecated, the megafaunal dispersal syndrome is implicated.
13. **The seeds germinate and grow well in upland habitats where planted, but the species almost exclusively inhabits floodplains in the wild.**[19] If large animal dispersers are no longer present, then gravity and flowing water fill in as the primary dispersal pathways. These abiotic pathways limit the species to floodplains.
14. **The geographic range is inexplicably patchy or restricted.** A species suffering dispersal problems is likely to experience range constriction. Moderate dispersal problems may lead to patchiness, with individuals or

> populations relatively isolated, despite the wider presence of suitable habitat. Severe problems may eventually lead to highly local endemism as the lineage plummets toward extinction. For example, mammal-dispersed species of the taxon Crescentieae (which includes the jicaro fruit, *Crescentia*, studied by Dan Janzen) may be restricted to a single mountain.[20]

In an odd twist, the megafaunal dispersal syndrome was first identified in a place with a dearth of megafauna: Costa Rica. Thirteen thousand years ago, at the end of the Pleistocene epoch, the Western Hemisphere lost three-fourths of its genera of mammals deer-sized or larger. The giant megaherbivores (the elephantids, ground sloths, glyptodonts, and notoungulates) vanished entirely. In the Americas, therefore, the megafaunal dispersal syndrome applies to fruits that are all, to some extent, anachronisms. But in Africa, where, biologically speaking, the Pleistocene never ended, a megafaunal dispersal syndrome should be functional as well as evident. We can thus look to the characteristics of still-functional megafaunal fruits in order to discern whether candidate anachronisms halfway around the world do indeed look, feel, smell, and taste the same.

African and Asian Analogs of American Anachronisms

In their 1982 paper, Janzen and Martin cited work by D.-Y. Alexandre, who had identified three classes of fruits dispersed only or primarily by African elephants:

1. Fruits 5 cm in diameter or larger, containing (usually) one seed protected by a very hard covering (like the stone of a peach).
2. Very big fruits, 10 cm or longer, containing multiple seeds that may be hard but are not protected by stony coverings.
3. Big woody pods containing an aromatic pulp and multiple seeds. Many of these pods require so much chewing that the lineages were forced to invest in formidable seed coats—so formidable that the seeds cannot absorb the water necessary for germination unless they are abraded.[21]

In the years since Alexandre offered his classification, and especially following publication of Janzen and Martin's proposed "megafaunal dispersal syndrome," a number of helpful studies of Old World fruits have been conducted. Takakazu Yumoto and Tamaki Maruhashi identified seven species of rainforest fruits in Zaire for which elephants are not only good dispersal agents but seemingly the only ones.[22] Gorillas join elephants as

MEGAFAUNAL FRUITS OF THE OLD AND NEW WORLD TROPICS. *Top:* ASIAN FRUITS—In the bowl, clockwise from top, are grapefruit, kiwi (native to China, not New Zealand), mango, tangerine, pomegranate, and carambola starfruit. To the right are sliced specimens. An extracted mango seed is alongside the sliced mango. *Bottom:* NEOTROPICAL FRUITS—In the bowl, clockwise from top, are avocado, guava (two), papaya (two), cherimoya, and passion fruit. The two specimens in the center are canistel, also known as egg fruit or yellow sapote. All of these tropical or subtropical fruits were purchased in specialty fruit stores in New York City.

vital dispersal agents for another two fruiting species. Among the elephant fruits they identified were *Tetrapleura tetraptera* (family Leguminosae) and *Omphalocarpum mortehani* (family Sapotaceae). *Tetrapleura tetraptera* is a pod 15 cm long with a high sugar content that, despite its sweet attractions, is deemed poisonous by local people. *Omphalocarpum*, with a rind so thick and hard that only elephants can crack it to get at the pulp, is similar to one of Janzen and Martin's prime examples of neotropical anachronisms: *Crescentia*.

Lee J. T. White, whose study site is in Gabon, notes that while most fruits consumed by elephants are taken after they have fallen, species on smaller trees and shrubs are often "delicately plucked with the trunk."[23] His group found, too, that of the seventy-two species of seeds defecated by elephants, only two species are consistently crushed. They concluded that plants dispersed by elephants generally have large, dull fruits with succulent flesh and a strong, "often yeasty" smell.

In yet another wide-ranging survey of elephant eating habits in Gabon, François Feer reports that the size of whole fruits consumed by elephants ranges from 2 cm to 27 cm in their greatest dimension.[24] The biggest seed found in elephant dung during this study was that of *Balanites wilsoniana*, which is 5 cm in diameter, but Feer notes that previous researchers found whole seeds as large as 10 cm in diameter (from *Borassus aithiopicus*, the palm family) in elephant dung. Elephants prefer fruits with fibrous pulp, and a third of the fruit species they consume are enveloped in a tough woody rind (like *Omphalocarpum procerum*). Tough-rinded fruits, along with very large fruits containing small seeds, are often squashed with a foot. Seed ejection sometimes occurs after chewing has detached seed from fibrous pulp. Intriguingly, various species of *Raphia* (in the palm family, Arecaceae) bear fruits containing a single large and very hard seed that so easily detaches from the surrounding pulp that elephants regularly spit it out. One wonders whether *Raphia* fruits are anachronistic, having evolved to attract some less discriminating beast.

Of the fifty-three species of seeds Feer found in elephant dung, only twelve were occasionally in fragments. Forest elephants in his study area of Gabon appear to be exclusive dispersers of ten species. He concluded that "large size is the common trait of fruit selected by elephants" and that his findings generally corroborate Janzen and Martin's criteria for the megafaunal dispersal syndrome.

Janzen and Martin 1982 is also cited in several reports that pertain to single fruit species consumed by Old World megafauna. One such study was undertaken not in Africa but in lowland Nepal. The authors, Eric Dinerstein and Chris M. Wemmer, present their work primarily as an

FRUIT DISPERSED BY RHINOCEROS IN NEPAL. *Trewia nudiflora* depends on rhinos for dispersal of its seed. The dried plant on this herbarium sheet was collected in 1921 and the fruits in 1927 and 1932.

effort to accumulate the kind of data about Eastern Hemisphere fruits essential for evaluating the validity of the megafaunal dispersal syndrome and the corollary anachronism concept.[25] The subject of their work is the fruit of *Trewia nudiflora* (family Euphorbiaceae); henceforth, trewia. Each tree produces about seventy fruits, which ripen and fall over the course of four months during the yearly monsoon. Trewia fruit is not especially large, with a diameter of 4 cm or less. What makes it megafaunal is its unpalatability to any animal (such as a carnivore) incapable of ingesting foods that are well defended chemically, along with the vulnerability of its seeds to abuse by small herbivores. Each fruit contains three seeds almost 1 cm in diameter. The flesh has the hardness and texture of a raw potato and tastes extremely bitter to the human palate, even when fully ripe.

Ripe trewia fruits are ignored not only by humans but also by monkeys, bats, and birds. Only large mammals seem interested in the pulp, and of these only the largest defecate viable seeds. There are no wild elephants in Royal Chitwan National Park of lowland Nepal, so the primary dispersal agent for trewia is the greater one-horned rhinoceros, *Rhinoceros unicornis*. Captive rhinos eagerly gobble three hundred trewia fruits in an hour, but wild rhinos are thought to consume on average two hundred fruits daily during the peak of the fruiting season. Where rhinos are present, *Trewia nudiflora* may constitute almost half of the trees in the lowland forest. Where rhinos have been extirpated, trewia is generally rare.

Another sign that rhinos are important for dispersal of trewia in Nepalese forests pertains to timing. Seeds that are not freed from the pulp by transit through an animal's gut do not have a chance to germinate before the end of the monsoon, which is the optimal germination and growing season. Also, trewia is shade intolerant; seedlings that germinate under the parent tree are doomed.

The strangest facet of the trewia-rhino association stems from the rhino habit of defecating in latrines (latrine is, in fact, the technical word). Rhinos move to well-used sites in their personal territories to defecate. Active latrines are elliptical, averaging seven meters in length, and they foster intense competition. Rhinoceros latrines frequently become luxuriant mats of trewia seedlings.

Dinerstein and Wemmer speculate that rhinos were important in the evolution of trewia fruit traits. Rhinocerotidae is one of the most ancient families of megafauna alive today. Thirty million years ago, the rhino family produced the largest land mammal of all time, an Asian rhino that stood five meters at the shoulder. The ecologists conclude their trewia report with the big-picture implications: "We believe that the most likely explanation for the occurrence of large indehiscent fruits in the Earth's woody flora is that these fruits, like *Trewia,* evolved in response to big herbivores. In Chitwan, woody plants producing such fruits constitute a relatively small proportion of the local flora. Nevertheless, the short species list, from a taxonomist's perspective, should not obscure its significance to the ecologist." They encourage similar studies elsewhere, especially in the tropical forests of Southeast Asia.

The enormous jackfruit of India and Malaysia (family Moraceae) and the ghastly-smelling but heavenly-tasting durian fruit of Southeast Asia (genus *Durio,* family Bombacaceae) have much to teach us about the Asian manifestation of the megafaunal dispersal syndrome. For these and other Asian fruits, published information on seed dispersal by animal mutualists is sparse and anecdotal.[26] Anecdote is nevertheless sufficient for me to confidently declare mango a megafaunal fruit. Although everybody loves the pulp of mango (genus *Mangifera*), only the biggest of the big will swallow the seed. Asian elephants, widely reported to have "a fondness" for mango, do not wait for the fruits to fall; they shake them out of the trees. Mango seedlings commonly sprout from elephant dung in northern Thailand. The Sumatran rhino has also been recorded as an enthusiastic harvester of fallen mango fruit. Like elephants, rhinos defecate the seed intact.[27]

An African fruit that depends on forest elephants for dispersal. *Balanites wilsoniana* contains a pit that only an elephant will swallow. Courtesy of Colin Chapman.

The best-studied Old World fruits that depend on megafauna for dispersal grow in the forests of central Africa. At an ecological research site in western Uganda, zoologists Lauren and Colin Chapman were intrigued by the huge seeds of *Balanites wilsoniana* (hereafter, balanites) that they saw poking out of elephant dung. They launched a study to determine whether the species does in fact depend on elephants.[28]

"The elephants have to swallow a lot of ballast for the nutrition they get out of it," Colin Chapman told me during an interview at the University of Florida. "Even so, balanites is one of their preferred fruits."

"I think the fruit of balanites has the most distinctive odor in the forest," Lauren Chapman added. Her office is across the hall from her husband's. "You can smell balanites a long distance away. It's like fermented apples—a nauseatingly sweet smell."

Together the Chapmans pieced together a description for me: Balanites is a forest tree of the upper canopy. The entire population produces fruits more or less synchronously each year, which drop upon ripening over a period of two months. Balanites fruit is shaped like a mango, averaging 9 cm in length and 6 cm in diameter. The fruit is protected by a thin green skin, like that of an avocado, and it browns upon ripening. Beneath is a layer of pulp, just a half centimeter thick, surrounding one very big seed. When ripe, the pulp has the brilliant green color and soft texture of kiwi fruit. The seed is protected by a pit with a convoluted surface like the pit of a peach, but much bigger, thicker, and more fibrous. The pulp clings to the convolutions with almost as much tenacity as the pulp of a mango clings to its pit. Seed spitting is therefore not an option for elephants.

The Chapmans did not know if balanites pulp tastes as sweet as it smells. They were understandably reluctant to perform such a test. The pulp contains diosgenin, a powerful steroid, making it poisonous to humans and other primates. Locally it has the reputation of killing any human who might consume it in quantity. Villagers do, however, make use of the pulp. They apply it to the scalp to kill lice and drop it into pools to stun fish. Because the poison is in the dose, great bulk is in the elephant's favor. During the two months of fruit fall, fourteen percent of elephant dung piles were found to contain balanites seeds. Of these, the average number of seeds per heap was ten. "That's not a big quantity," Lauren Chapman commented. Even when fruits are plentiful, elephants continue to feed overwhelmingly on high-fiber plant materials—leaves, bark, and grasses.

Elephants lack the front-end detoxifying vats of ruminant mammals such as antelopes and giraffes. Elephants, rhinos, and zebras all house symbiotic microbes in their hindgut instead. Hindgut microbes may be just as effective as forestomach microbes in breaking down toxins, but by the time bacteria in the large intestine of an elephant have a chance to perform this service, poisons already have infiltrated the animal's bloodstream through the lining of the small intestine. Elephants therefore depend on a behavioral trait rather than purely physiological tools to detoxify much of their food: they eat clay.[29] The Ugandan elephants studied by the Chapmans regularly ate a red clay "very similar in composition to the pharmaceutical Kaopectate."[30] Whether by instinct or tradition, these elephants can eat balanites fruits because they engage in a behavior that overcivilized humans (mis)regard as unseemly for themselves. (We will revisit this point in Chapter 8.)

The team concluded that forest elephants in Africa not only aid dispersal of balanites seeds but are mandatory. The seed is too big to be swallowed by any other forest animal, and balanites has little or no recourse to backup dispersers. Several species of squirrels, along with bush pigs and insects, do prey upon the seeds, but there is no evidence of transport and hoarding. Squirrels do not consume the whole seed, however; they extract just the embryo. The Chapmans do not know why squirrels are so selective. Perhaps the rodents are seeking nutrients concentrated in the embryo. More likely, the rest of the seed—the energy-rich endosperm—is too heavily laced with toxins for a mammal to ingest. Some rodents who hoard seeds use another trick to avoid poisons: they consume the seed only after it germinates. Germination deactivates many toxins, as Asian peoples who consume bean sprouts long ago discovered.[31]

Rodents are thus no help to the balanites lineage. Elephants are the only effective dispersers. In fact, elephants do such an excellent job of dispersing balanites seeds that it is reasonable to think of their kind as the target disperser for which evolution shaped and flavored balanites fruits. Most of the seeds not swallowed by elephants will suffer rodent or fungal attack. The Chapmans found that when the pulp rots, so does the seed. Fungal hyphae surround and penetrate the pit.

Elephants are vital for more than preventing seed rot and moving progeny beyond the confines of the parent tree. Passage through an elephant enhances germination. In a test planting, the Chapmans compared germination success of 232 seeds collected from elephant dung with 457 control seeds removed from freshly fallen fruit. Half of the dung seeds germinated, but fewer than one percent of the controls did. At the conclusion of the experiment, the Chapmans opened ungerminated seeds to determine whether perhaps the controls simply needed more time. The answer was no. Virtually all of the embryos were visibly dead.

Balanites wilsoniana is thus an ultimate megafaunal fruit. It evolved to attract the services of elephants and is now utterly dependent upon them.

We can use the traits of balanites and other megafaunal fruits of Africa and Asia to identify candidate anachronisms in the Americas. But can a suspected anachronism ever be moved beyond the status of candidate? Can we gather more compelling evidence than fruit form and dispersal distress? One obvious path would be to test-feed New World fruits to Old World megafauna. I describe preliminary results of such tests on zoo and other captive elephants in Chapter 8. Captive elephants are not the only prospects for experimentation, however. At least two candidate anachronisms of the New World are now available to wild elephants in Africa: avocado and papaya. These have become popular orchard crops in tropical lands throughout the world. Nevertheless, only one paper, published in 1993, provides any data (and cursory at that) on elephant interest in these exotic fruits. Perhaps because they are not native, these fruits are judged unworthy of attention in field studies of plant-animal interactions.

Martin N. Tchamba and Prosper M. Seme studied the diet and feeding behavior of elephants in a forest reserve of Cameroon. More than half of the elephant dung piles they examined contained seeds of one or more of twenty-two species of fruiting plant. Their study of African forest elephants in a seemingly wild setting was disrupted in a way fortuitous to those of us interested in New World anachronisms. Because the reserve was so small (and growing smaller, owing to intrusion by desperate peasants clearing land for coffee cultivation), the elephants were forced to forage beyond its borders. The result is that dung samples contained not only indigenous seeds but also the seeds of three exotic fruits: the Asian mango (family Anacardiaceae), the American papaya (family Caricaceae), and the American avocado (family Lauraceae). The contention that papaya and avocado are megafaunal anachronisms thus received a substantial boost.

Though small-seeded, *Carica papaya* qualifies as an elephant fruit in two ways. First and foremost, this small tree bears big fruits: 10 cm long in the wild, with domesticated varieties bred to far larger dimensions. The location of the small seeds (smaller in wild populations than in cultivars) is the second suspect trait. Unlike figs, whose tiny seeds are scattered throughout the pulp, papaya seeds are clustered at the center. I found I could easily crush papaya seeds between my molars, but one experiment was enough. They have a pungent, peppery taste. Mammals that pillage papaya pulp are thus unlikely to ingest many seeds; we simply eat around the seedy center. Although insect repulsion and phylogenetic inertia may explain the placement of papaya seeds, the fitness of this lineage must owe to millions of years of interactions with beasts too big to eat around them. Because the pulp is readily mashed between tongue and palate, the delicate seeds should not have been subjected to the molar mills of target dispersers. The ghosts of evolution that haunt papaya orchards today probably include South America's extinct toxodons, ground sloths, and gomphotheres.

One more Old World study requires mention. In a sad but illuminating twist, A. Gautier-Hion and colleagues cite Janzen and Martin's work for this reason: The tropical rainforest in Gabon has become so depleted of large mammals (including elephants) that current dispersers of fruits bearing megafaunal traits were mostly "food-hoarding rodents, for which the mesocarp [pulp] is simply a barrier to be removed" in quest of the seed.[32] Elephant fruits even in Africa can become *locally* anachronistic, overbuilt for the backup dispersers on which they now depend.

A worse fate awaits balanites. Local extirpation of elephants can be expected to precipitate the local extirpation of balanites. The fruit becomes an anachronism with no backup crew. In a study site where elephants are rare, the Chapman team observed balanites seedlings growing only under the canopies of parent trees. Yet where elephants are common, "a good proportion of balanites seedlings were found some distance from large conspecific trees." The Chapmans have been told of a forest in Kenya that purportedly contains only mature balanites trees; no young trees or seedlings are evident. That forest has been bereft of elephants for thirty years.

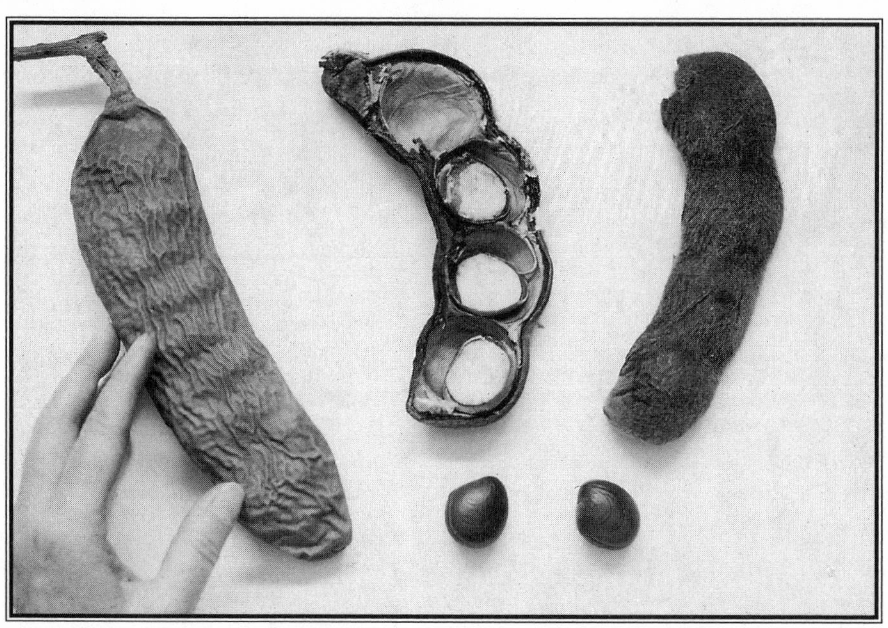

A NEOTROPICAL POD FORMERLY MISUNDERSTOOD. In 1971 Dan Janzen concluded that the seeds of *Dioclea megacarpa,* a canopy-level vine of climax forests, were "dispersed away from the vine by arboreal and terrestrial rodents." Janzen has since reinterpreted this fruit as a megafaunal anachronism. *D. megacarpa* is at left, with sister species *D. violaceae* at right. The three seeds in the open pod have been sliced in half.

4

Advancing the Theory

DAN JANZEN AND PAUL MARTIN'S CLAIM THAT MANY LARGE NEOTROPICAL fruits are anachronisms has given rise to remarkably little controversy. A paucity of criticism can, of course, mean that an idea is so unworthy that nobody wants to waste time refuting it. This has not been the case with neotropical anachronisms. Many studies favorably cite Janzen and Martin's 1982 paper, and some take it as a central thesis. Even so, to the extent that anachronisms have been integrated into ecological and evolutionary thinking, they turn up as isolated curiosities rather than standard and expected components of biotic communities in many regions of the globe.

How significant are anachronisms? How much of an ecological presence do they have? Mauro Galetti, Carlos Yamashita, and Paulo Guimares are cataloguing fruit diameters in the forests and savannas of Brazil, with the aim of distinguishing "megafauna dependent" from "megafauna independent" fruits. Whether Brazil's largest native mammal—the tapir—willingly ingests a fruit and defecates viable seeds will provide a second test of megafaunal dependence.

That kind of detailed, quantitative study would be useful, too, elsewhere in South America, especially if coupled with surveys of ecological indicators such as rotting fruit, confinement of a species to floodplains, or increased populations in the vicinity of horses and cattle.[1] A fruit-by-fruit quantitative and natural history study would also be useful for the thirty-some Costa Rican plants Janzen and Martin offer as anachronisms. Part of that work has been done—by Dan Janzen or his close associates. The Costa Rican fruits that Janzen had studied in some detail before his anachronism insight, and that thus required fresh interpretation afterward, included the palm fruit *Scheelea rostrata,* the fleshy fruit *Guazuma ulmifolia,* and six legume pods, among them two featured in the first two chapters of this book: *Enterolobium cyclocarpum* and *Cassia grandis.*[2]

What about other large fruits in Central America? Or in Mexico or the Caribbean? Might there be anachronisms in Madagascar, where giant lemurs and flightless birds bigger than ostriches roamed until the coming of humans? Researchers have, in fact, speculated that some of the largest fruits on the island of Madagascar may indeed be megafaunal anachronisms.[3] How about New Zealand, Australia, Japan, China? Might we find megafaunal fruits made *locally* anachronistic in India or Southeast Asia, and thus receding in range, wherever the countryside has been purged of elephants and rhinos? Have botanists and ecologists working in any of these places puzzled over the riddle of rotting fruit?

What about North America? What about right here where I live—the temperate deciduous forests of the eastern United States and the desert scrub and woodland of the American Southwest? I launched this book project expecting to find the work had been done, the answers right at hand. It hasn't and they aren't. No scientist has conducted even a first-pass evaluation of the most obvious U.S. prospects.

Now that I know what to look for, however, I don't have to look far. Candidate anachronisms jump out of the landscape. Important natural history details of each of these prospects have already been published, though not with the aim of deciding the anachronism question. More, these fruits are readily available for me to pluck, probe, sniff, and taste. Some I have been able to observe rotting where they fall.

This chapter and the next three present that first-pass survey. Candidate anachronisms from North America are scrutinized one by one, as I have already done for honey locust. Anachronistic features of persimmon, pawpaw, osage orange, and Kentucky coffee tree are evaluated for eastern forest environments and gourd, cactus, devil's claw, mesquite, and wild tomatoes for the desert West. We'll take a look at anachronistic thorns, too—following the lead of Janzen and Martin, who proposed many such defensive candidates.

Before beginning the survey of anachronisms in my neighborhoods, we shall consider one in-depth critique of the concept and how that challenge not only can be answered but can advance our understanding of anachronism theory. First, however, let us turn our attention to a particular fruit-animal pairing on a remote island in the Indian Ocean. Work published on this curiosity in 1977 preceded the full-blown research program initiated by Janzen and Martin. The debate this early paper generated shows how a grasp of the larger pattern trumps the minutiae of isolated examples.

The Dodo Controversy

The tambalacoque tree (or tam; *Calvaria major*) is found only on the island of Mauritius in the Indian Ocean, about a thousand kilometers east of Madagascar. This tall, buttressed tree of the tropical rainforest produces fruits the size and structure of a peach, with an outer surface that ripens, like balanites and avocado, from green to brown. The tam evolved within family Sapotaceae, which is rich in tropical fruits having functional megafaunal traits in the Old World and anachronistic megafaunal traits in the New.

There are no native megafauna on Mauritius or on the neighboring islands of Réunion and Rodrigues (the trio is known as the Mascarene Islands). Like the Galápagos Islands, the volcanically formed Mascarenes are too remote to be reached by any mammal set adrift on a raft of roots. But birds have no trouble getting there. The pigeon and rail families are especially good at devolving flightlessness from lineages that arrive on islands impoverished in ground-based predators. By abandoning flight, the birds open up foraging possibilities, favoring the evolution of a gut and gizzard that can digest bulky, fibrous fare. Once on this path, the now-grounded bird lineage can begin to accommodate the desire of tropical, closed-canopy forest trees to evolve seeds big enough to sustain seedlings for a long time on dimly lit forest floor. Trees can evolve big seeds only if somebody can be induced to swallow them.

The monkeys and pigs that live in the remnant forests of Mauritius today were introduced by colonizing humans during the past few hundred years. Monkeys and pigs, moreover, do more harm than good to big sapotaceous fruits. Monkeys are superb pulp thieves, and pigs are seed predators extraordinaire. The other sapotaceous fruits of the Sapotaceae family on Mauritius and neighboring islands are small enough to be consumed by resident, flighted fruit pigeons, yet tam fruit is startlingly large. The fruit is spherical, 5 to 6 cm across. The stony pit is almost as big and is blanketed by a pulp (that holds tightly to the pit) only a half centimeter thick. No bird now living on the island can swallow the seed. Nor would one want to, as indigestible ballast is antithetical to an airborne mode of life.

The tam tree did not float to Mauritius Island from Africa or Madagascar; it is thought to have evolved in place from a smaller-seeded species of *Calvaria* that probably hitched a ride inside a fruit pigeon or a swimming tortoise. The tam fruit evolved on Mauritius to please and to survive passage through the island's biggest flightless bird. The dodo, *Raphus cucullatus,* stood nearly a meter tall.

That was the hypothesis offered by ecologist Stanley Temple in 1977. His theory that the tam tree depended on the dodo for reproduction attracted attention for a number of reasons. First, Temple published the idea in the journal *Science.* Many people read *Science* on a regular basis

THE BIRD THAT HAUNTS AN ENDANGERED TREE ON MAURITIUS ISLAND. The dodo, clubbed to extinction by sailors just three hundred years ago, was a vital disperser of the now rare tambalacoque tree.

(I do), so many people encountered Temple's idea. Second, the scientific story behind the sad plight of a magnificent tree in an exotic land struck a chord and was widely reported in the media. The dodo is for many people the very symbol of human culpability in the worldwide decline of biodiversity. The slow, naive birds were easily clubbed to death by European sailors restocking ship larders. A Dutch explorer wrote in 1649, "We lived on tortoises, dodos, pigeons, turtle doves, gray parrots, and other game, which [the crew] caught by hand in the woods."[4] By the late 1600s, the dodos (and soon thereafter, the tortoises and gray parrots) were gone.

The dodo's story is poignant, too, because its strangeness was born of the whimsy of evolution shaping ecological interactions out of scraps of life swept onto a shore far from any other. An isolated island might well produce a creature who hasn't a chance against brutes honed by struggles of continental scale. Had we primates shown restraint in our harvest of the dodo, the pigs we set loose on Mauritius would have done the deed for us.

The dodo-tam idea also made an impact because Temple offered several bold claims. He stated that this was the first documented example of a plant threatened with extinction because an animal mutualist had expired. He reported that only thirteen tam trees survived on the entire island, and all were three hundred or more years old. The implication: no new trees had been recruited into the forest since the demise of the dodo.

The tam tree clearly was on its way out. The reason for its extreme rarity, Temple concluded, was the loss of the tree's obligatory dispersal partner. Animal assistance in reproduction of the tam tree is not just a matter of getting the seed out from under the canopy of a parent and away from pulp thieves and seed predators. In this case it includes preparing the pit for germination. Temple judged that once the tam had evolved a mutualism with the dodo, it could not do without. An exceedingly thick stony pit—thicker even than the layer of pulp surrounding it—protected the seed from the ravages of a dodo gizzard that, in turn, had evolved the means to crush fibrous material, including coated seeds. The dodo was primarily a seed predator, but the tam had found a way to transform predation into dispersal. Because the plant evolved seed protection hardy enough to survive such treatment, it now utterly depends on abrasion; a germinating seed could not break through its shield without such abuse—or so Temple thought.

Gizzards of big herbivorous birds are generally nastier than molars of big herbivorous mammals. A tough seed is subjected to molars for maybe ten seconds as an elephant or rhino or horse grinds its food. But taken into a gizzard, the seed might remain in a dangerous environment for hours, even days. There, in the company of other hard seeds, it is kneaded and knocked about by a muscular organ lined with biological sandpaper. Even worse, the gizzard is a mosh pit of stones the bird has swallowed for the express purpose of enhancing abrasion. A bird can afford to break an ingested stone on a too-hard seed; a mammal is a fool to risk a broken tooth.

In a two-step process, Temple tested his hypothesis that tam trees depended on the gizzard of a bird for successful reproduction. First, he force-fed tam pits to turkeys, whose size and feeding habits make them surrogates for dodos. Turkey gizzards are adept at battering down the fortresses surrounding seeds tough as hickory. Of seventeen tam pits Temple forced down turkey gullets, seven were crushed by the birds' gizzards, the seeds then easily digested. Of the ten pits that survived the ordeal (for as long as six days), some were regurgitated and the rest were sufficiently reduced to be defecated. All these survivors, however, had been abraded, some severely.

The second stage was the planting of survivors. Three of the ten germinated. Temple exulted, "These may well have been the first *Calvaria* seeds to germinate in more than 300 years." His report ends, "These observations provide empirical support for the hypothesis that the fruits of *Calvaria* had become highly specialized through coevolution with the dodo. After the dodo became extinct, no other animal on Mauritius was capable of ingesting the large pits. As a result, *C. major* has apparently been unable to reproduce for three hundred years and nearly became extinct. The findings presented in this report may provide a basis for preserving the species through propagation of artificially abraded seeds."

Five months after Temple published his idea, the first opposition appeared in print.[5] In a letter to the editor of *Science News*, one critic took issue with Temple's report that only thirteen trees remained in the wild; there were more, the author alleged, and some of the younger ones had surely got their start without the assistance of the dodo. This critic concluded that large-scale deforestation hurt the slow-growing tam tree far more than had loss of a putative dispersal partner. A second refutation appeared in *Science* in 1979. A. W. Owadally of the governmental forestry service on Mauritius criticized the factual content of Temple's paper. More, he offered an alternative hypothesis, that the decline of the tam tree owed to depredations of monkeys and the invasion of exotic plants. Was there a political motivation to this opposition? Owadally concluded, "It is necessary to dispel the tambalacoque-dodo 'myth' and recognize the efforts of the Forestry Service of Mauritius to propagate this magnificent tree of the upland plateau."

Temple's rejoinder, published in the same issue of *Science*, continued to argue the facts of the case, which were, necessarily, hard to pin down. The ages of the surviving wild tam trees could not be known with certainty because slow-growing tropical trees do not develop annual growth rings (from alternating periods of dormancy and growth) the way temperate trees do. Although no bones of dodo had been found in the vicinity of existing tam trees on the plateau, fossils of any kind are unusual in uplands; bones are preserved in basins where sediments accumulate. There was disagreement, too, over whether historical records showed tam trees growing in the lowlands before colonists razed those forests. Finally, Temple suggested that his critic had failed to report one crucial fact about the propagation of tam trees: "The Mauritius Forestry Service has only recently succeeded in propagating *Calvaria* seeds, and the unmentioned reason for their recent success strengthens the case for mutualism. Success was achieved when the seeds were mechanically abraded before planting."

The debate flared again in 1987, when A. S. Cheke added two more items to the list of possible reasons the tam tree was on the verge of extinction. Cheke proposed that deer introduced to the island were eating the seedlings. Also, the low rate of germination owed not to unabraded pits but to low rates of pollination. In other words, the fruits looked normal but they contained nonviable seeds. (One wonders, however, if diminished pollination was more a consequence than a cause of rarity.) Cheke's strongest counterargument was that Mauritian foresters had stopped abrading tam pits before planting because the practice was found to have made no difference to germination. Rather, tam pits "have a natural zone of weakness in the woody seed coat, where they always split on germination." Cheke's criticism edged toward ill will in his concluding paragraphs: "Clearly it requires more than a turkey and a few seeds to prove 'obligatory mutualism' between an extinct bird and a rare tree. . . . Future research must be devoted to conservation rather than to creating scientific folklore of a kind that . . . is extraordinarily hard to root out once established."

In an odd twist, Cheke wrote, "If any animal was necessary to the tambalacoque, was it not the tortoise?" Giant land tortoises originally from Madagascar had long ago colonized the Mascarene Islands, as tortoises are buoyant and swim readily. The only population in the whole oceanic region that survived human contact, however, was on the coral atoll of Aldabra, four hundred kilometers northwest of Madagascar. Another researcher had proposed in a 1978 paper that tortoises might have been the agents by which some plants spread from Madagascar to outlying islands.[6] In addition, the vital role tortoises play in the reproduction of a tomato native to the Galápagos Islands, *Lycopersicon esculentum*, had been established in 1960.[7] The tortoises help by moving the fruit about. More important, passage in the gut of a tortoise is required to break seed dormancy, which is otherwise "extreme and of indefinite duration."

It must be pointed out that the digestive system of a tortoise is gentler on seeds than that of a dodo. Reptiles lack a gizzard, and some ingested fruits pass through virtually intact. The thin case that protects the small seeds of a wild tomato is nothing like the bony pit surrounding a tam seed.

Cheke did propose one helpful action of the dodo in tam tree reproduction. "It may be relevant that Malayan species of *Calophyllum* and *Canarium* need to have the flesh of the fruit removed by animals before they will germinate." The importance of pulp removal (rather than pit abrasion) for the tam tree was, in fact, the conclusion drawn in 1988 by a trio of botanists. Peter Jackson and colleagues suggested, too, that another sapotaceous fruit on the decline in Mauritius, *Mimuseps petiolaris*, is suffering not from unabraded pits but from rotten seeds. Unless the pulp is removed by something, fungal hyphae rotting the pulp will work their way into the pit. Extinct parrots, an extinct giant skink, an extinct coconut crab, two species of extinct tortoises, and the dodo were offered as prospective pulp thieves.

The importance of pulp removal to prevent seed rot is mentioned in a variety of papers on seed dispersal.[8] Janzen and Martin 1982 is among them. "In horse-free habitats the indehiscent fruits [of *Crescentia alata*] lie on the ground and rot in the rainy season, and fermentation of the fruit pulp kills the seeds." The pulp-removal hypothesis for the tam tree received support in 1991 in a lengthy review article published in the journal *Oikos*. Mark Witmer expressed frustration that Temple's dodo-tam mutualism had become "textbook dogma," despite opposition. Criticizing Temple's turkey experiments for very low sample size and lack of a control, Witmer also refuted Temple's report that only thirteen exceedingly old tam trees were still alive in the wild. The actual number, he contended, was several hundred. He further concluded that because the dodo was not a specialist frugivore but primarily a seed and fiber consumer, dodos probably destroyed most of the tam seeds they consumed. The dodo was therefore just one of a suite of potential dispersers that included tortoises, skinks, and parrots. Introduced monkeys and pigs were no substitute because "they may not clean seeds adequately and they destroy seed and seedlings of many endemic plants."

Pointing out that other fruits with thick and bony pits—peach, cherry, walnut, and hickory—germinate just fine without abrasion, Witmer surmised that pulp removal is the only action crucial for germination. "Extinct frugivores were no doubt critical in cleaning and dispersing tambalacoque and other endemic Mauritian forest tree seeds." Why, then, does the plant produce such a formidable pit? Witmer's answer: "It is likely that the tambalacoque evolved a thick, tough seed coat in response to consumption by its dispersers, but there is no evidence that the seeds require abrasion before they can germinate."

The dodo-tam hypothesis no longer generates heated debate in scientific journals, though not because the question has been settled.[9] It is just that there is little more to say, other than speculation. The issue becomes one expert's opinion against another's. Historical questions are sometimes unresolvable.

It is not my place to enter an opinion either, and I shall not choose an expert for a final say-so. Whether or not the tam tree depended on the dodo for seed dispersal or germination is beside the point. We need only look at the fruit traits of the tam tree and how its current coterie of dispersers handles the crop to sniff out an anachronism. Tam fruit is unquestionably overbuilt and suffers the consequences. If the critics are right that the only animal assistance the fruit requires is removal of the pulp, not abrasion of the pit, then why create attractive pulp at all? If nobody is swallowing the fruit whole (and that fact is apparently undisputed), then the fruit is too large for current gullets.

It doesn't matter whether a dodo, a tortoise, or even a giant skink was doing the swallowing. Whether the pit required abrasion or merely needed to survive it is not the issue either. Any way you look at it, *Calvaria major* produces fruit that is anachronistic in its native home today.

Then, too, that tam trees are not entirely incapable of reproducing in the wild makes little difference from an anachronism point of view. It matters not whether the surviving trees number thirteen or a hundred, whether they are centuries or mere decades old. Either way, the tam tree is having trouble. If plants instantly went extinct when their *prime* dispersal method ceased functioning, Dan Janzen would not have stumbled across rotting fruits in Costa Rica and Paul Martin would not have conjured Pleistocene ghosts. A species need not be an utter failure at reproduction, bereft of secondary dispersers, in order for its fruit to be deemed an anachronism. Again, the fruits must simply be construed as overbuilt for current conditions; something about them suggests a waste of precious energy; one trait or another puts them out of step with their ecological surrounds. And then a reasonable scenario must be developed of a bygone time in which those same traits would have been useful rather than extraneous or harmful.

Which, after all, is the more prudent course in science? When ascertaining the evolutionary ecology of an organism, do we limit our reach to the present, no matter how impoverished the present may be? Or do we welcome ideas that include the whole cast of characters that accompanied a lineage like the tam tree or honey locust in its evolutionary journey?

Degrees of Anachronism

The most important and only thorough critique offered in response to Janzen and Martin's 1982 paper, "Neotropical Anachronisms: The Fruits the Gomphotheres Ate," was published three years later. "Gomphothere Fruits: A Critique" was Henry Howe's challenge. Howe is a prominent ecologist who, like Janzen, has conducted fieldwork (including seed dispersal studies) in the neotropics. His review appeared in 1985 in *American Naturalist,* one of the most prestigious periodicals in ecology. Counterarguments to the anachronism hypothesis were thus aired in a forum widely read by ecologists, botanists, and zoologists.

The review was mixed. Because he dressed the negative points in dramatic language, the positive points are easy to miss on a quick read. Nevertheless, in his abstract and summary, Howe's assessment was balanced and restrained. He judged the hypothesis of megafaunal anachronisms to be "intriguing" yet "vague and difficult to reconcile with ecological theory and practice." It has "a ring of truth" and is thus "worth a searching review. It is hoped that an edifying refinement will evolve from the turmoil." In the summary paragraph, Howe pronounced the idea "not yet a useful tool."

Sandwiched between these judicious conclusions one finds fire. In the body of the paper, Howe charges that "consideration of unknown history risks dragging an entire discipline into the mire of untestable assertions, circular thinking, and ad hoc reasoning." Howe depicts the anachronism idea as adaptationist thinking taken to an extreme and criticizes Janzen and Martin for advocating that "imaginary pristine communities of the past" become the standard by which dispersal inefficiencies of the present would be judged.

Dan Janzen anticipated this core criticism. In a book chapter that appeared in 1979, he introduced the anachronism idea with passion: "Evolutionary biologists are very fond either of pretending that the plant traits we see are selected for and maintained by current interactions or, at the least, of choosing to work on those systems that seem to match this assumption. However, we all know perfectly well that a plant (and its herbivores) is a collection of anachronistic traits that at any given time have caught up with contemporary selective pressures to a highly variable degree. One reason why evolutionary biologists like to sweep this fact under the rug is that for a long time it was a standard loophole for dealing with some conspicuous trait, the natural history of which was not known. . . . To invoke currently extinct selective pressures to explain the presence of a trait, however, was to mask an incomplete study of natural history. Another reason for avoiding the anachronistic aspects of evolutionary biology is that one is caught in a morass of untestable hypotheses."[10]

Henry Howe backed his criticisms of "untestable assertions" with specific grievances. He was wary of the ecological criteria that Janzen and Martin used to distinguish the megafaunal dispersal syndrome in places, like the neotropics, where megafauna are extinct. Yet it was an ecological observation, "the riddle of the rotting fruit," that had launched Janzen on this path of inquiry.

Two years before the anachronism idea appeared in *Science*, Howe had written of "the enormous seed waste" of a Panamanian rainforest tree, *Tetragastris panamensis*, and of "the 'inefficiency' of the dominant dispersal agents."[11] About the same time, he reported the ecology of seed dispersal for the wild nutmeg *(Virola surinamensis)* in Panama, concluding that the small assemblage of bird and mammal dispersal agents for this tree was "anomalous."[12] For neither of these puzzles did Howe deem it necessary to reach into the past for a ghostly explanation. The puzzles simply remained puzzles. Perhaps they still do. Both of these large fruits are bright red and open (dehisce) to dangle seeds individually coated with fleshy aril that attracts monkeys and large birds. Under no stretch could these fruits be deemed megafaunal, and yet they too often rot on the forest floor. Before we give up on finding a paleoecological explanation for the riddle of the rotting red fruits, however, we should keep in mind that a monkey more than twice the size of the largest monkey alive in the neotropics today accompanied the ground sloths into extinction.[13]

Lauren and Colin Chapman, the ecologists who determined that African balanites fruit depends upon elephants for dispersal, find merit on both sides of the issue. Colin told me, "I really liked Howe's paper, the rebuttal. He brought up all the negative points. Most of what Howe said is: How are you going to prove this? Which is true. We need creative ways of getting around Howe's criticisms. Blindly accepting Janzen and Martin's idea wouldn't be good either."

He continued, "I think it's a fun hypothesis, but there hasn't been a lot of work done on it yet. So I'm still just being a little cautious. Even so, it's a really great idea."

Lauren added, "It's a nice template for sprouting ideas, but I would be careful about putting a tag on every fruit—"

"Without carefully looking at the alternative hypotheses," Colin finished.

For example, drawing on their experience in the forests of Uganda, the Chapmans do not necessarily perceive a riddle in rotting fruit. Colin explained, "Elephants do not always find the trees." Lauren elaborated, "There's so much rotting fruit on the ground. Balanites trees can be forty meters tall. They're one of the taller canopy trees, and they have a big crown, so the fruit crop is big and it is spread over a pretty wide area. So even when the elephants are visiting the tree, there's a lot of fruit left on the ground."

If the elephant population had not been depleted by hunting, I inquired, would there still be fruit rotting on the ground?

Lauren was first to respond. "I would guess so. The tree's strategy may be to produce a really big fruit crop to attract elephants. If some is left over, well, that's the cost of attracting a big group of elephants."

Any grand new idea is going to contain errors. To prompt others (not to mention oneself) to consider giving up entrenched ways of thinking, the originators of the idea will probably overreach. Most of Howe's critique of

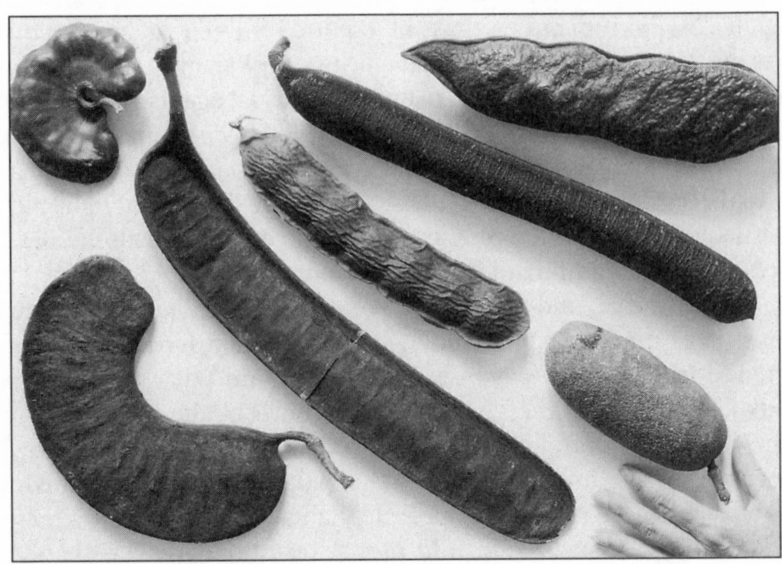

COMPELLING ANACHRONISMS. The harshest critic of anachronism theory nevertheless judged that "tough, indehiscent legumes may be the most obvious candidates for the syndrome." The seven species of tropical pods shown here were collected in Brazil, Peru, and Nicaragua.

the anachronism paper is about overreaching. I recall that my own first reading of Janzen and Martin's list of thirty-seven "native trees and large shrubs . . . whose seeds were probably dispersed by extinct megafauna" brought a furrow to my brow at the entry that reads "*Ficus* spp." Figs have tiny seeds, and the fruit is not encased in a husk that would fend off birds, bats, and monkeys. Janzen himself had written in 1979, "Who eats figs? Everybody. Wild figs are famous for being consumed by a very large number of species of vertebrates."[14] My gut reaction was that figs ought not be in Janzen and Martin's list of Costa Rican anachronisms. Howe reacted the same way, and so did a team of scientists who studied seed dispersal in Gabon.[15]

The problem for Howe was that the list of Costa Rican fruits and the list of megafaunal fruit traits offered by Janzen and Martin were too broad, generating too many exceptions. The idea was vague and not yet useful, he alleged, because the megafaunal syndrome as then depicted "lacks consistent morphological criteria" that could be applied in a practical way. Some of the fruits were soft; some were hard. Some persisted on the tree when ripe; others were shed. Some were large, and some not so large—and how large is "large" anyway? Figs were not acceptable to Howe, no matter how many may rot on the forest floor. Still, "Tough, indehiscent legumes may be the most obvious candidates for the syndrome," he wrote.

Later, Howe gave his approval of a pared-down, reined-in version of the megafaunal dispersal syndrome. He allowed that "this hypothesis may have a core of truth for plants with highly resistant seeds and seedlings. . . . At least half of the species listed by Janzen and Martin seem to conform to other syndromes, and in fact have effective living dispersal agents in protected forests. Just the same, this intriguing hypothesis deserves continuing attention as more becomes known about undisputed 'megafaunal' plants in Africa and Asia."[16]

The idea, Howe urged, should be pursued on a fruit-by-fruit basis. He speculated that "some neotropical fruits, such as that of *Gustavia superba* . . . might once have been dispersed by large mammals that became extinct during the Pleistocene."[17] His wariness of the broad reach of the syndrome as originally defined stemmed from his wariness of other fashionable dispersal syndromes. "Syndromes remain a useful ecological taxonomy," he wrote, "but [are] not a substitute for actual study of the dispersal process and its consequences." Later he reiterated, "Syndromes identify general patterns that serve as paradigms for more discriminating analyses."[18]

Dan Janzen seems to agree. In email correspondence Janzen told me he was not interested in "the pseudoformality" of seed dispersal syndromes. What is important is that "one looks at a present-day fruit (or seed) and simply asks, to what degree could its traits have been selected for and maintained by what we see around us, and to what degree would it have been shaped by extinct selective agents?"

I don't see that a preference for one approach cancels out the value of the other. Dispersal syndromes and single-fruit analyses are complementary at different scales of perception. Janzen and Martin 1982 is a synergy of the two perspectives. Janzen contributed the examples of individual fruits; Martin set those examples in the context of a megafaunal dispersal syndrome. Similarly, I don't see how the inherent vagueness of and exceptions to the megafaunal dispersal syndrome cancel out its value for screening a whole flora for candidate anachronisms—each of which should then be studied in detail. Ecology is, after all, a science of exceptions.

Henry Howe wrote approvingly of the anachronism idea once again in his 1988 book, *Ecological Relationships of Plants and Animals*. Here he departed from his earliest critique by accepting as evidence of anachronism an ecological indicator: fruit rotting beneath the tree. Nevertheless, in the three venues in which he reflected on the anachronism idea and the megafaunal dispersal syndrome, Howe never failed to point out the peril of overreaching. And he persisted in raising one other troubling issue: If a plant has, in fact, lost its cadre of dispersal agents, then it "probably could not survive for 500 years without dispersal, much less ten millennia."[19] He notes, too, that African trees deprived of elephant dispersers begin to disappear within decades.[20]

The persistence problem deserves attention. Janzen has written, "Considering that the megafauna was extinguished and that not all trees whose fruits were dispersed by it also disappeared, there is the puzzle of what did disperse the seeds of these trees during the last 10,000 years. The remaining large contemporary herbivores are obvious candidates."[21] The biggest of the native stock are three species of tapir in Central and South America. Indeed, tapirs have recently been depicted as "the last representatives of the neotropical Pleistocene megafauna."[22]

Janzen studied the dung of baird's tapir *(Tapirus bairdii)* in Costa Rica and discovered that the tapir spits, fragments, or digestively kills the seeds of many species of fruits with megafaunal characteristics. He reports: "The large sweet indehiscent fruits of *Guazuma ulmifolia, Pithecellobium saman, Enterolobium cyclocarpum, Caesalpinia coriari, Crescentia alata, Manikara zapota,* and *Mastochodendron capriri* all appear to have evolved in response to large mammals as seed dispersal agents. However, the latter two species have seeds that are too large and too weak to withstand the molar mill of a tapir, and their seeds were probably originally dispersed by larger mammals that chewed their food even less thoroughly, or at least chewed fruits even less thoroughly. In addition, their disperser coterie has probably always contained an array of small mammals as well [hoarding animals], and it was probably these animals that kept these trees in the game after the Pleistocene extinction (assuming that some dispersal is necessary for a population to persist)."[23]

By observing a captive animal, Janzen learned that tapir spit out virtually every seed of the legume *Hymenaea courbaril,* that they digestively destroy all seeds of *Cassia grandis,* and that they spit out half of the seeds in a guanacaste pod and digest three-fourths of the rest. Even if a tapir defecates in fine form the seeds of a particular tree, a tapir population is not well equipped to handle the whole harvest.[24] "Everything in moderation" seems to be their guiding principle. Janzen's captive tapir would accept only two to ten fruits of any one species in a meal, waiting to be offered fruits of something else and then something else. The animal accepted leafy browse in a similar manner. Janzen concluded that for many plants the tapir is a "trivial" disperser. "At this rate it would take years for armies of tapirs to eat the thousands of fruits that fall beneath a large adult guanacaste tree."[25]

José Fragoso and Jean Huffman disagree. Studying tapirs in the wild twenty years after Janzen's observations of a captive tapir, they found many big seeds in tapir dung, including a species of *Cassia.* As the last of the Pleistocene megafaunal lot, tapirs may now be vital dispersers of big-seeded neotropical trees. Fragoso and Huffman judged that tapirs "play a critical role in the structuring of tropical forests."[26]

What about Henry Howe's central charge that the anachronism idea is vague and overextended? One small but important restatement of anachronism theory offers a solution: we must forgo the binary concept and begin treating anachronism as a continuum.

Until now, the thinking implied by Janzen and Martin and assumed by Howe (and others, I have discovered) has been that something either is or is not an anachronism.[27] Either the fruit of avocado is an anachronism or it is not. Either the guanacaste tree is missing its partners in propagation or it is not. Janzen and Martin's 1982 presentation of the anachronism idea may have fostered this assumption—although that was apparently not their intent.[28] Misinterpretation has nevertheless generated unnecessary discord. Henceforth, we should talk about *degrees of anachronism*. Some fruits are more anachronistic than others.

The root of this solution is prevalent in Janzen's work. Consider his judgment about tapir: "Tapir is a seed predator to some trees, a potential good dispersal agent to others, and a mixed blessing to yet others."[29] I asked Janzen how he felt about an emphasis on the gradational aspect of anachronisms. He responded, "Since all organisms are buckets of anachronisms of various ages, obviously there are degrees of anachronism for any trait and any organism. The central point is that evolutionary processes do not—repeat, do not—cause a given organism to be a perfect mirror of its current or possible selective processes."

I propose a three-part system of grading: *moderate anachronisms, substantial anachronisms*, and *extreme anachronisms*. This system works well for the North American anachronisms that are the focus of this book, but another way of distinguishing degrees may be preferable elsewhere. In this three-part system, only the extreme category would demand that no native animal of any kind be engaged in even casual dispersal of the fruit, which nevertheless bears the traits of an animal dispersal syndrome. The fruit must rot where it falls. But it must rot not because the fruit is so abundant that available dispersers become satiated. It must rot not because overproduction is the strategy for luring dispersers. It must rot because nobody finds the pulp the least bit attractive. In floodplains, extreme anachronisms may be moved by rising or running water, but the fruit should nevertheless appear overbuilt for that function. There should be cheaper ways to build buoyancy, given the lineage's developmental constraints.

Plants that produce extremely anachronistic fruits will have made major investments in pulp that not only is wasted but endangers the seeds when fungal hyphae attack. The pulp is so noxious or inaccessible that not even

DEGREES OF ANACHRONISM IN NORTH AMERICAN FRUITS. LEFT: Moderate anachronisms (mesquite, cactus, persimmon); MIDDLE: Substantial anachronisms (devil's claw, pawpaw, desert gourd, and honey locust); RIGHT: Extreme anachronisms (Kentucky coffee tree, osage orange). One seed is alongside each fruit.

pulp thieves take an interest. It is not unusual for fruits dispersed by mammals to endow their *seeds* with potent toxins to discourage grinding. But if everybody shuns the part of the fruit that is supposed to serve as bait, then somebody must be missing. Except for a proviso that insect deterrence surely plays a role in toxic pulp, one can safely pronounce the fruit overbuilt if the pulp is *too* toxic and serves no dispersal function commensurate with its cost of construction.

When Henry Howe objected to Janzen and Martin's characterization of some neotropical fruits as megafaunal anachronisms, he may have believed that for a fruit to qualify as an anachronism it must possess the kinds of morphological traits and offer the sorts of ecological indicators that I here define as extreme. Nobody must disperse the seeds—anytime, anywhere. Similarly, Stanley Temple's portrayal of the dodo-tambalacoque relationship could not stand up to the exacting demands of critics who believed they had demolished Temple's scenario by showing that extant creatures do serve as dispersers on occasion and that some trees had got their start centuries after the dodo became a ghost.

Extreme anachronisms constitute the most compelling category. They toss up the most insistent puzzles. But lesser degrees of dispersal inefficiencies, too, deserve the name *anachronism*. These lesser degrees differ from the extreme in that some dispersal is effected. The fruit is nevertheless overbuilt for current modes.

For substantial anachronisms, there are dispersers, but no animal (at least none whose population and appetite are up to the task) can be regarded as the *target* dispersal agent; all are haphazard, less effectual mutualists.[30] Pulp thieves, seed predators, or both will disperse seeds of substantial anachronisms only inefficiently. To use Dan Janzen's preferred terminology, the "seed shadow" will be restricted and patchy. The poorly matched interactors will rarely or never assist plant reproduction by defecating or regurgitating viable seeds. Dispersal happens not because a seed is swallowed but because a thief runs off with a fruit or a seed-containing fragment of fruit in order to consume the pulp in a safer place. The fruit will be disassembled at the new site, the thief dropping or spitting virtually all seeds. Likewise, a seed predator will occasionally hoard a fruit or the extracted seeds for later eating, and some of the stores will never be retrieved. Dispersal therefore will happen, but by a mode very different from that which shaped the fruit's evolution.

Finally, those in the weakest category, moderate anachronisms, are not lacking dispersal mutualists, but something about these interactions seems out of kilter. The fruit is overbuilt in some way or is seriously overabundant, suggesting that the plants once counted on beasts with bigger appetites to show up for the harvest.

Among the species dealt with in depth thus far, guanacaste pods of Costa Rica and honey locust pods of the eastern United States should probably be considered substantial anachronisms. Introduced livestock are the only effective dispersal agents for these fruits, which otherwise rot where they fall. The smaller and easily crushable mesquite pods would be moderate anachronisms. Mesquite was dispersed before horses and cattle invaded the landscape, but this desert tree is dispersed much more capably now.

Among the fleshy fruits, avocado is, at minimum, a substantial anachronism. Avocado suffers from a lot of pulp thieves, but nobody except the rare jaguar will swallow seed.[31] As for extreme anachronisms, two were introduced in Chapter 1: osage orange and ginkgo. Their stories and that of another are told in Chapter 6.

This approach to anachronisms does not allow judgment to be made on an assessment of fruit traits alone. The kind, quality, and extent of interactions with animals are crucial considerations. In this way Henry Howe was right. There will always be exceptions to any list of traits and ecological indicators construed for anachronisms. There will continue to be a place for close examination of the natural history of plants on a species-by-species basis, as Dan Janzen has done for a number of Costa Rican trees. Nevertheless, the generalizations pioneered by Paul Martin will help field scientists determine which candidates are worth investigating.

Why Pawpaw Grows in Patches

One more distinction among the grades is important. Especially if an anachronism is thought to be extreme, and if its missing partners are believed to have vanished with the Pleistocene, how did the species survive so long without them? This is the question that drove Henry Howe's skepticism.

There is fossil evidence of only one plant extinction (a spruce) occurring since the late Pleistocene.[32] Fossil botanical evidence of any kind (and of fruits in particular) is hard to come by, however, especially in the tropics. Plants dependent on megafauna may have gone extinct, but there is no hard evidence. Nevertheless, we have abundant evidence of range shrinkage—of extinctions in the making. Several relatives of *Crescentia alata,* for example, have geographic ranges today limited to just a few valleys or even a single mountain.[33] Dan Janzen maintains that scattered local extinctions would have been inevitable following the disappearance of dispersal partners, but lineages nevertheless hung on in places "where dispersal is not such an important part of their biology."[34]

More generally, the answer to Henry Howe's challenge will turn on *compensatory life history traits.* If a fruit is rarely, if ever, dispersed by animals, then the plant had better have extraordinary skills for keeping the lineage going in other ways. The story of the pawpaw is a good place to begin this investigation.

The pawpaw tree *(Asimina triloba)* bears the largest edible fleshy fruit of any tree native to the United States. Big fleshy fruits in the Western Hemisphere are otherwise limited to tropical or subtropical habitats. Pawpaw is the only species of family Annonaceae that survives not only frosts but many months of solid winter. Because pawpaw is an understory tree of temperate deciduous forests, and because it produces large seeds with large stores of energy, it has evolved an ability to sprout from a seed and establish in the shade of the parent. This is a fortuitous trait for a plant whose dispersal is no longer assured.

Pawpaw grows wild in twenty-five eastern states, as far north as southern Michigan, where I grew up. As a child I heard about pawpaw (mostly by way of the folk song "Way Down Yonder in the Pawpaw Patch"), but I never encountered the tree or its fruit. I still haven't in the wild, though I have been able to find it in botanical gardens. If you live near a pawpaw patch, you've probably seen (maybe even eaten) the strange fruit. If you don't, you are probably as clueless as I was about what they look, smell, and taste like.

Whether and to what degree pawpaw fruit qualifies as an ecological anachronism is dealt with in the next chapter. For now, characteristics of the tree itself provide the model for understanding the tricks of persistence. Primary is the capacity for individuals of the species to live for a very long time. If the life expectancy of an individual tree is, say, a thousand years, then only thirteen generations would have elapsed since the end of the Pleistocene. Each generation would have a thousand cycles of spring, summer, and autumn to grow a fresh crop of fruit. If during those thousand years, even one seed is launched into the world in a way that yields a fruit-bearing tree that in turn reproduces successfully, the lineage will persist. The scenario sounds plausible for sequoia, but what about a tree like pawpaw that rarely develops a trunk wider than fifteen centimeters? A pawpaw tree becomes an elder around age twenty-five. For a single pawpaw trunk (technically, a stem) to survive half a century is almost unknown. The part of the pawpaw we see—the stem and branches—is short-lived by tree standards. Pawpaw's longevity resides, rather, in its root.

Many trees send up new, tentative shoots from the base of the main stem after the above-ground part of the tree has been injured or killed. Some species sprout stems-in-waiting while the current growth is still vigorous. One or more of these nascent stems will shoot up to produce a new tree if the current one fails or is severed by a storm. A select group of trees, however, can sprout suckers from shallow, laterally spreading roots (called root runners) extending meters beyond the main stem. Some species will establish new stems by this method while the main stem is still healthy, and all will produce an army of fresh sprouts when the elder sickens or dies. Because the new sprouts emerge from the same rootstock, the replacement stems are actually not new plants. The stems are new, but the genotype—the individual organism—is the same. The ability of some trees to "coppice" in this manner is what for thousands of years has provided villagers in Europe with a local, renewable source of wood for cooking and heating and leafy branches to feed livestock.

Root runners make it possible for one seedling to become a thicket of stems occupying a large area. Pawpaw excels in this strategy. One pawpaw tree can become a vast patch. How old might a clonal patch of pawpaw be? Nobody has looked into the matter. Until now, there has been no reason to pose the question. Clones of aspen in the Rocky Mountains have been reported as old as eleven thousand years—essentially, as old as ice-free terrain has been available. A clone of huckleberry growing on a ridge in Pennsylvania has been estimated at thirteen thousand years. Expanding clumps of creosote bush in the arid lands of the western United States may be almost as ancient.[35] It is possible, just possible, that a pawpaw patch in a Kentucky woodland may be underlain by a network of roots that thirteen thousand years ago first pressed downward through a heap of mastodon dung.[36]

One-fifth of the nontropical trees native to North America—all of them angiosperms—have some capacity for root sprouting, though many (in contrast to pawpaw) do so only after their main stem has been injured or killed.[37] Root suckering is more pronounced in understory trees, such as pawpaw, than in trees of the forest canopy. Pawpaw surely did not evolve cloning skills in order to survive the loss of its prime dispersal agents. That's not the way evolution works. The lineage must have developed cloning capacities for some other reason millions of years ago. Propagation by root runners is, in fact, "a regular trait in some primary understory trees."[38] Small trees that make do beneath a dense forest canopy may have a difficult time getting enough sunlight to establish from a seed. More assured is to have photosynthates supplied by elder clonemates until a fresh stem can feed itself.

That ability to sprout from lateral roots comes in handy today. The lineage was "preadapted" to cope with dispersal problems. Long-term ecological conditions thought to promote vegetative forms of reproduction are cold or nutrient-poor habitats as well as excessive shade.[39] Habitats prone to fire, storm, or flood damage would also select for cloning talents.[40] Clonal traits should be favored, too, in habitats subject to heavy browsing or grazing, such as forest edge.

Anyone who has ever seeded a new lawn and watched the spattering of tender stems become a plush carpet has witnessed the power of cloning. Especially for lawns fastidiously mowed, there is little opportunity for seeds to develop. Cloning is the only way for most bare patches of lawn to fill in.

Along the Gila River in New Mexico, I practice a little forestry on about ten acres of floodplain. Cottonwood trees (same genus as aspen, *Populus*) and willows *(Salix)* exhibit awesome powers to resprout from their roots after a flood or a beaver demolishes the stems. I've also been astonished at how far from the "parent" stem a new cottonwood sucker might emerge on a terrace of sand, far too high above the water line for the stem to have started from seed. Another telltale sign of a sprout from root rather than seed is speed of growth, as the new stem is well fed by clonemates. Often, too, the sprouts from root runners will have abnormally large leaves.

How good is pawpaw at cloning? Stunningly capable—it has to be. Inadequate seed dispersal is not the tree's only sexual dysfunction. Inadequate pollination threatens the species as well. One research team observed only a few instances of insect visitation to several patches of blooming wild pawpaw in their study area in central Illinois.[41] Each patch contained between 62 and 196 (presumably clonal) stems. Does the paucity of pawpaw pollinators signal that the flower is anachronistic too?

The flower of pawpaw mimics carrion or dung. Petals are brownish maroon, and the blossom is reported to give off a fetid or yeasty odor. In mid-May I spent two hours watching and photographing at least four species of fly that were visiting flowers in the little pawpaw patch at the Brooklyn Botanical Garden. I could not detect a floral odor, but the flies obviously could. In contrast to the Illinois findings, here the flies were prolific. Yellow grains of pollen were easy to spot on the hairy backs of the larger insects. But is the back of a fly the intended vessel for pollen transport?

The flies I observed all entered a flower by walking along the inner surfaces of the petals, their backs scraping the sides of the reproductive dome at the flower center, which bears receptive female stigmas at its tip and (several days later) pollen-bearing male stamens along the flanks. At the base of the petals is the reward: a corrugated surface that exudes a liquid imbibed by flies. There can be no question that the back of a large, hairy fly captures and holds pollen. But can the back of a fly transfer pollen onto a receptive female stigma?

A 1992 study of a Florida species of pawpaw, *Asimina parviflora*, reports that even though flies are the most frequent floral visitors, beetles may be the effective pollinators.[42] Tropical members of family Annonaceae are known to be pollinated by beetles, especially large scarab beetles—a.k.a. dung beetles. Effective pollination by flies is rare for these plants, and I imagine the reason is the backside problem. Flies may be able to hang onto smooth petals, but beetles cannot. Thus the downward-facing flower of pawpaw may be accessible to a heavy beetle only if the beetle uses the reproductive dome for traction—landing on the stigmatic tip and dragging its belly along the anthers as it crawls upward toward the floral reward.

In commercial pawpaw orchards, pollination is usually achieved by hand. Adding to the plant's problems, pollen produced by anthers of the same genotype may have difficulty fertilizing ova in the same patch, no matter how many stems there may be. The flowers in a clone are to some degree self-incompatible. If the next pawpaw patch is a mile away (which is not unusual in recent times; suggestive of a species in decline?), beetles must make the long traverse before losing the dusting of pollen they received in the first patch.

Patchiness is a feature of pawpaw ecology in two ways. Pawpaw tends to grow in clonal groups, or patches, and its distribution is patchy within its range. Where pawpaw is common, one should suspect a history of human intervention. The earliest Europeans exploring the eastern part of the continent reported that Native Americans practiced pawpaw horticulture—which leads us to another compensatory life history trait that may

AN ANACHRONISTIC FLOWER? Might pawpaw's pollination difficulties owe to extinction of beetles that had depended on megafaunal carrion or dung? TOP: A fly enters a downward-hanging pawpaw flower by nimbly walking on the petals, attracted by the fetid odor and carrion color. The hole near the base of the petal may be where tissue was consumed by a pollinator or by a nectar thief that excavated from the outside, perhaps before the bud had opened. BOTTOM: The anatomy of the flower indicates that flies may be ineffectual pollinators. Pollen scraped onto the hairs of a fly's back from the central reproductive dome may rarely transfer to the female organs of another flower.

be found in suspected Pleistocene anachronisms: human mutualists. Humans should be regarded as pulp thieves of pawpaw because we do not swallow the seeds, but we more than compensate for such poor behavior by purposefully propagating the plant. Even before horticulture was practiced, early humans must have carried the fruits from forest to camp, effectively dispersing the seed while stealing the pulp. Thus, if some plants lost their prime dispersal partners at the end of the Pleistocene, those that appealed to humans immediately gained a replacement.[43]

Humans occupying North America lost a great deal of their effectiveness as dispersal agents for pawpaw when they began importing bananas and other tropical fruits, which were, moreover, available year-round. The short season of pawpaw harvest was no longer an event to be anticipated. Pawpaws were also driven into decline wherever wild woods were transformed into subdivisions. Even if pawpaws are planted intentionally in the new human-built landscape, the plants are doomed if the surrounds are mowed.

I witnessed that pawpaw peril at the New York Botanical Garden, which is in the Bronx, a long subway ride from the Brooklyn Botanical Garden where I had spied on flies. The trees in the Bronx were a disappointment; they had almost as many dead branches as living. The problem? It seems they had been denied their patch.

There were only two pawpaw stems, about two meters apart. The bigger stem had a trunk diameter of about 12 cm; the other, 8 cm. In other words, they were verging on elderly, and it was time for the root system to replace them. Surely the roots had tried and tried again, foiled by whirling blades; the two trees were surrounded by mowed lawn. Four meters away in a bed of mayapples was a single, robust young stem. Having invaded the designated territory of another species, the newcomer risked discovery and removal. I offered the staff advice for saving the pawpaw: stop mowing! Permit the pawpaw to express its true nature. Let the pawpaw specimens become a pawpaw patch.

As it turns out, staff at the New York Botanical Garden are well aware of the pawpaw's "problem."[44] The concern is not so much that these specimens will die but that the root suckers will invade everything else in this part of the garden if given the chance. Pawpaw is, I am told, allowed to sucker with abandon in the patch that thrives in the sector of the garden that is managed as wild woodland.

The ability to coppice and to send forth root runners that establish new stems may thus be saving features for plants whose fruits become anachronistic. Among the North American anachronisms discussed thus far, honey locust and mesquite are excellent coppicers. Is it perhaps more than coincidence that all of the candidate anachronisms to be discussed in the next two chapters are skilled in the art of cloning?

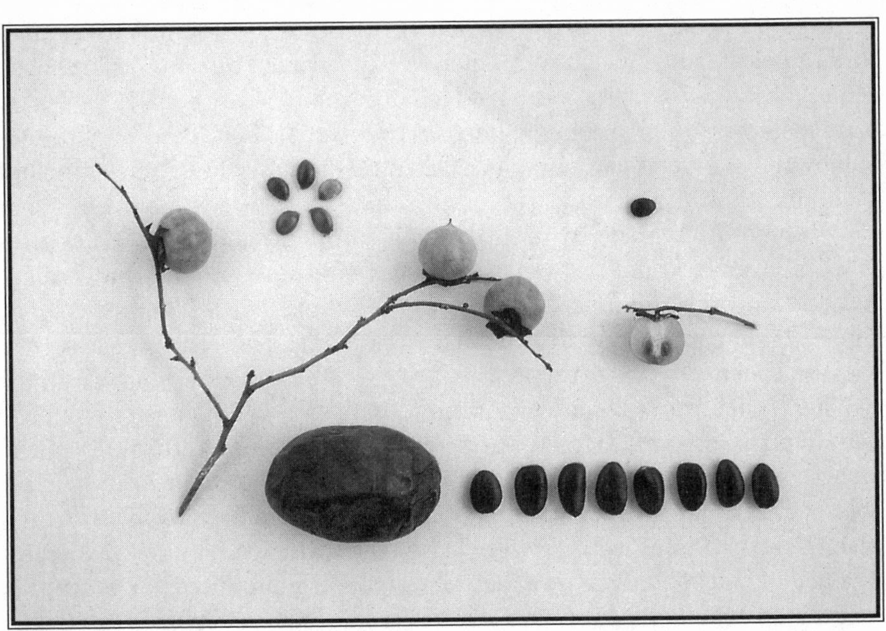

The two largest pulpy fruits native to temperate North America. A single overripe fruit of pawpaw *(Asimina triloba)* is shown with underripe persimmon fruits *(Diospyros virginiana)* still on the branch. Honey locust seed (1 cm) for scale.

5

A Fruitful Longing

ON MY DINNER PLATE THIS EVENING WERE TWO FRUITS. For the sliced tomato, I played the role of a good dispersal agent. Its small seeds were barely noticeable in the pulp, which I mashed more than chewed before swallowing. The seeds will come out the other end in fine shape, I surmise, though I have no plans to formally test this hypothesis.[1] The other fruit on my plate is misrepresented as a vegetable even more often than is the tomato. It is a designer squash so sweet and tender that I served it naked of butter or honey.

Had I been pressed for time, I might have treated the squash like a pulp thief, tossing the uncooked seeds onto the compost pile. There they would have a chance of sprouting in the spring. More likely they would be scavenged by a larder hoarder, the rock squirrel who lives nearby and stashes all sorts of seeds for lean times. But this evening I decided to become a seed predator too. While the pulp was steaming on the stove, I put the seeds into the oven to roast.

Earlier today I established yet another kind of relationship with a plant. I carried an apple up a nearby arroyo and ate it after topping the canyon rim. Save for the fact that I left the core in a habitat so dry the seeds haven't a chance, I should qualify as a serendipitously helpful pulp thief. I dropped the core along the unmarked trail, which is rarely traveled by a biped other than me. By now a quadruped, fox or coyote most likely, has surely discovered the treat and wolfed it down, cyanide-laced seeds and all.

I hadn't been on that trail in over a month. A lot of shit accumulated during my absence. In the course of maybe half a kilometer, I came upon nine distinct leavings of fox and one complex contribution from a coyote. This time of year, fox dung contains almost exclusively the seeds and indigestible fiber of juniper berries. (Technically, juniper, like all conifers and other gymnosperms, does not produce fruit. Only flowering plants do. *Diaspore,* same root as *diaspora,* is the more precise term. Many botanists, however, simply remind the reader of the distinction and then revert to the more familiar term. So shall I.)

Given this wealth of juniper-rich fox excrement, I dissected only the sample that was fresh enough to attract flies. As usual, there was not a trace of rodent bone or hair. The oil-rich fruits of juniper trees seem to provide the foxes with all they need, so the rodents get an autumnal break.

For several weeks, I had been confused about the content of such dung. Juniper had crossed my mind, but as few berries were still on the trees and none visible on the ground, I had set that option aside.

"Juniper," declared my neighbor, Allen Campbell, who has been in this canyon all his life.

"No shit!"

"The reason you don't see any on the ground," he continued, "is because they've all been eaten. Foxes and coyotes love 'em."

A week earlier I had left on his doorstep a plastic bag with a sample, asking for an opinion. "Wash off the rind and I bet it will still smell like juniper," he suggested. "You could crunch a seed just to be sure."

In addition to fox feces, the animal trail where I placed the apple core was marked by a monster of a coyote dropping, in three parts. This scat contained four distinct food items and a plug of grass. (Grass is not a food for canids; it is a cleanser.) One segment was made entirely of fine, soft fur. Another segment contained burnished red fragments of a huge scorpionlike arthropod, known locally as a vinegarone, along with seeds of juniper and cactus. The fragments of cactus skin were the color of a manila folder, so I knew it was cholla and not the purple fruit of prickly pear. A few paces down the trail a cholla still bore a single, lemon yellow fruit. The fruit was beyond ripe and beginning to desiccate. Such fruit must not be a preferred food for canids, as this one was within reach of a coyote and, with some effort, a fox.

Tomorrow I will scour our yard and our neighbor's looking for a particular kind of fox feces. I hope to find a sample that contains the pits of wild persimmon fruits, *Diospyros virginiana*. We are a long way from persimmon country, but I know that turd exists. Three days ago, I fed a half dozen ripe fruits of wild persimmon (obtained by mail from Indiana) to my friend Mrs. Foxie. Today I fed her four more, while carefully observing how she ate them. I wanted to know whether foxes were pulp thieves of persimmon or legitimate dispersal agents.

Mrs. Foxie is a gray fox, *Urocyon cinereoargenteus*. Her personal name may be a bit precious for a serious book, but I didn't name her for this experiment. She became Mrs. Foxie when I began to recognize her as an individual four years ago. For five years before that, I had been on friendly terms with her mate, Foxie. Toward the end of the last summer that we helped Foxie feed his kits, he became lame. I found a fox skeleton in an arroyo near the house the following year. During Foxie's final summer, a female would sometimes lurk in the brush when he came to visit. Who else could she be but Mrs. Foxie? The following summer Mrs. Foxie took her former mate's place, sitting doggie style in precisely his spot, five paces from the porch steps, waiting for a handout. The two notches in her left ear were unmistakable.

Jane Goodall opened up serious science to the use of names rather than numbers when animals are observed as individuals. Dan Janzen used names—Tinto and Negrito—for the horses that helped him ascertain the megafaunal attractions of guanacaste pods. Those horses already had names, as it is customary for ranchers to name their horses. When Janzen conducted the same study on a few cows he rented for the experiment, he used numbers. Ranchers do not name their cattle, though dairy farmers often do.

Persimmons Were Not Made for Possums

Five days have passed since Mrs. Foxie swallowed the last of the persimmons. I've scouted the grounds and a good distance along the road. I have come upon a lot of shit, but none bearing evidence of persimmon. I cannot, therefore, report conclusively that our gray fox defecated intact seeds, but I can surmise that she did. For one thing, I watched carefully as she ate the fruits. Sitting on the ground, I tossed the persimmons one at a time to a point less than a meter away. Each fruit was small enough for her to take into her mouth whole, chew a few times, and swallow. Only once did orange pulp spill out of her mouth while chewing (perhaps she was trying to rid the piece of its woody calyx), but she retrieved it. I examined the ground afterward and found one scrap of flesh, clinging to a calyx.

The fruits of wild persimmon in the eastern United States contain anywhere from one to eight seeds. Those eaten by Mrs. Foxie had five. Wild persimmons can be as big as five centimeters in diameter. The fruits Mrs. Foxie ate were about half that, much smaller than the persimmons in grocery stores. Cultivated persimmons were bred from an east Asian species closely related to North America's, and generation upon generation has been selected for traits valued by humans. Some members of *Diospyros* will produce fruits even when pollen is unavailable, resulting in seedless fruits. That trait has been exploited for commercial purposes. Persimmons sold in stores are seedless.

GRAY FOX SNIFFING A PERSIMMON. Although she had never before encountered persimmon fruit around her New Mexico home, this fox eagerly consumed a half dozen wild samples imported from Indiana.

Mrs. Foxie did not spit out the seeds for a very good reason. Had she attempted to do so, she would have lost much of the pulp. Ripe fruits of *Diospyros virginiana* bind the succulent pulp tightly to the seed. Even our own agile tongues cannot work the seeds against our teeth so as to completely separate the pulp. Everybody I know who eats wild persimmons spits out the seeds. Some remove the skin, but I am not that fastidious. The seeds do come away easily from unripe fruits, but nobody would want to eat a persimmon in that condition, as the tannins are intensely astringent—which is, in fact, why they are there. A dispersal agent is no better than a pulp thief if it takes fruit before the seeds are prepared for an independent life.

One early writer reported that an unripe persimmon "shrivels your mouth painfully so that you'd rather be stung by a wasp." Another wrote that the fruit "is so puckery it almost strangles one." Captain John Smith, of the first European settlement in Jamestown, Virginia, warned, "If it be not ripe, it will draw a man's mouth awry with much torment."[2] Not that adventurous, I sampled a mildly unripe persimmon. The experience is much like tasting a green banana. One's mouth feels coated in chalk.

Mrs. Foxie swallowed the seeds along with the pulp of persimmons, but did she swallow them whole, as a good disperser should? I surmise that

Evidence of the dispersal prowess of foxes. Clockwise from top: Gray fox dung with cantaloupe and squash seeds; dung with seeds of cultivated Concord grape; dung with apple seeds and large fragments of apple skin; suspected banana skin extracted from a sample not shown; dung with opuntia cactus seeds; dung with juniper seeds. Honey locust seed (1 cm) for scale at center.

she did. She made no crunching sound and chewed only a few times before swallowing. To crush a persimmon seed would take a lot more effort. I have tried without success to fragment one, using household pliers. Only a serrated knife would work.

Another reason I know she did not fragment the seeds is that she is still alive and well. Persimmon pulp is nutritious and delicious, but its seeds are highly toxic. Indiana lore holds that if you happen to swallow a seed and it "explodes" inside you, you're as good as dead. That may be an exaggeration, but there is a kernel of truth to it. Seeds are often toxic, as plants want their progeny to be swallowed whole, not chewed. Seeds are, after all, rich in nutrients that must be chemically or physically fenced off from the hungry world.

Perhaps the most compelling evidence that Mrs. Foxie defecated persimmon seeds intact can be found in my collection of fox feces. Pictured here are samples of fox dung that contain not only intact seeds of juniper but also apple, grape, cactus, cantaloupe, and squash. The squash seed is almost as big as the seed of persimmon, and it is far easier for us primates to crush.

American persimmon is a small tree of deciduous forests, its trunk rarely exceeding 20 cm in diameter. Its native range stretches from southern Connecticut to southern Florida and as far west as eastern Kansas and Oklahoma. Slow-growing, with very hard wood, the tree tolerates shade in the forest understory but readily invades abandoned fields. Worldwide, the two hundred species of persimmon (genus *Diospyros*) are almost exclusively tropical. One tropical species endemic to the island of Mauritius is about as endangered as a plant can get: only a single tree remains.[3] The persimmon family, Ebenaceae, shares its tropical preference with the family that pawpaw hails from: Annonaceae. There is no question that the richest region for anachronistic fruits is the American tropics, but I can nevertheless find enough prospects here in the temperate North to provide case studies for examining a range of matters pertinent to ecological anachronisms everywhere.

Is the American persimmon anachronistic? If so, to what degree? Persimmon was one of five fruits Dan Janzen and Paul Martin proposed as candidate anachronisms of temperate North America. All five (and one more) are examined in this chapter and the next.

Because an animal native to North America, the gray fox, readily eats persimmon fruits and swallows the seeds, I can conclusively state that American persimmon is not an extreme anachronism. In fact, persimmon may not be much of an anachronism at all. It may be missing the biggest and most voracious of its longtime mutualists, but the crew that remains is still swallowing seeds and defecating them in a viable state. If persimmon is anachronistic, it is only moderately so—the weakest category in the continuum.

To begin the inquiry, we need to learn whether anybody besides fox serves as a legitimate dispersal agent for persimmon. Does anybody else swallow and defecate seeds whole?

Raccoons do. In fact, raccoons *(Procyon lotor)* are exceedingly fond of persimmon. An experiment using captive animals revealed that, of nine foods offered, persimmon was consistently preferred. Raccoons favor persimmons over corn, earthworms, crayfish, and even chicken eggs.[4]

Black bears *(Ursus americanus)* are fabulous dispersers of seeds. They love fruits and they rarely fragment the seeds.[5] Bears are so fond of persimmon that they don't wait for the fruits to fall.[6] They climb trees to begin the harvest—which brings up a crucial point: persimmon fruits ripen in any one region over the course of several months. Even when ripe, a fruit may linger on the tree (and in good condition) for a month or more. Fruits will thus be available for consumption over a long time. Large animals with large appetites and a propensity to roam with the seasons may not be necessary to ensure that the fruit does not go to waste. Small critters that live locally, like foxes and raccoons—even box turtles are known to disperse persimmon—may be able to handle the load week by week.[7] What is more, persimmon fruits can linger on the ground for an awfully long time before succumbing to molds and fungi. I found one lone fruit beneath a persimmon tree in Florida on December 30. It was still delicious.

In mid-February hundreds of persimmon fruits lay beneath a single specimen of *Diospyros lotus* in the Brooklyn Botanical Garden. I had gone to the garden in quest of honey locust and holly but stumbled on this persimmon species from western Asia. The crocodile-skin bark (nearly identical to the bark of American persimmon) caught my eye, then the carpet of blue-black fruit. Raccoons and foxes are in short supply in Brooklyn; in this intensely urban environment, food on the ground had lain untouched for months. Though wrinkled, the fruits looked palatable. The brown pulp sparkled with crystals of sugar, like dried fig, and tasted great.

Fossil evidence of persimmon in North America extends back at least nine million years, into the Miocene. As I saw in the botanical garden, the species of persimmon on this continent today is very similar to the species in temperate Asia. Because persimmons with the same pulpy attractions occur globally, it is a good guess that the design suits a range of potential animal mutualists. American persimmon may have planned on a wide range of mutualists but perhaps not the cloven-hoofed clan, which depends on thriving colonies of microbes in forestomachs to digest leafy browse. (Recall that Dan Janzen found that cattle would not eat the sugar-rich pulp of *Crescentia*.) The literature is conflicted on whether white-tail deer, which are ruminants, consume the temperate species of persimmon in eastern woodlands of the United States. Even if they do, they probably snack on a few fruits at a time, so as not to disrupt rumen alkalinity.

Thus far we have seen that foxes, raccoons, and bears are legitimate and eager dispersal agents. Persimmons may have evolved with such frugivorous carnivores in mind. Indeed, Carlos Herrera believes there exists a carnivore dispersal syndrome for some plants—juniper fruits being one example. But he warns that members of this guild today represent only "the vanishing remnants" of a more diverse group of mammals that dispersed such fruits before the end-Pleistocene extinction.[8] There are, however, two problems with carnivores as seed dispersal agents. One is that carnivores are not as common as herbivores of the same size. Medium-size carnivores, such as foxes and raccoons, are probably more numerous today in persimmon habitat than they were five hundred years ago. The guns that extirpated wolves and cougars and bobcats brought forth an explosion of foxes and raccoons, whose primary enemies now are feral

OPOSSUM IN A PERSIMMON TREE. John James Audubon chose the persimmon tree as the setting for his painting of America's opossum, *Didelphis virginiana*. Nevertheless, persimmon was not made for possum.

dogs, steel traps, and highways. This phenomenon has been called "mesopredator release," and it is why we must be cautious about interpreting the adequacy of seed dispersal by midsize carnivores today as indicative of their role in the past.

A second reason that carnivores are less than perfect dispersal agents is that some of them place a great deal of value on their own excrement. Feces are not something to be dropped just anywhere; they are important markers of territory. A rock outcrop or the middle of a busy trail makes an ideal spot to broadcast one's presence—but a lousy place for a seed to grow into a tree. These are not "safe sites."[9] For this reason, foxes have been judged "legitimate but inefficient" dispersal agents.[10] Foxes may work in a pinch, but perhaps persimmon did not evolve with foxlike omnivores in mind.

I can attest to the propensity of foxes to deposit their treasures in unsafe sites. The reason fox dung has been so easy to collect is that I know where to look. Our compost heap is a valuable chunk of territory, and the amount of fox dung that shows up there proves it. Mrs. Foxie's new mate is bent on claiming even our porch steps. Two weeks ago I stepped out of the trailer to get something from the shed. The new Mr. Foxie was hanging out in the usual fox spot, but I ignored him. A few minutes later, I returned to find dung on the lowest stair of the porch.

One smaller carnivorous mammal must be evaluated as a potential disperser of American persimmon: the opossum. *Didelphis virginiana* is North America's only marsupial mammal. This lineage originated in South America, wandering northward only after the Isthmus of Panama joined the two continents about three million years ago. An avid climber, possum is found so often in persimmon trees when the fruit is ripe that the tree is known in some locales as "possum tree." John James Audubon, in his portfolio of North American mammals, chose to paint the possum feeding on fruits in a persimmon tree. There is no doubt that possums consume the pulp. But do they swallow the seed?

A monograph on the opossum published in 1945 concludes that possum dung previously reported as containing persimmon seeds must have been misidentified.[11] Eleven captive opossums, even when fed persimmons exclusively for a week, almost never swallowed a seed. Only one dung sample of sixty-three collected during that time contained a persimmon seed. This contrasts with the dung of two captive raccoons on a similar diet; all the raccoon scats contained persimmon seeds. For two reasons, then, I can say with conviction that persimmons were not made for possum. Opossums do not regularly swallow the seeds, and their kind was sequestered in South America for millions of years while the lineage of persimmon now in North America was taking shape on this continent or in Asia.

I asked Dan Janzen whether persimmon might be adapted for small carnivores more than megafauna. He replied, "A herd of mastodons would have had a great time vacuuming up *Diospyros* fruits after the first frost and taking them off the trees as well. A possum would have been lucky to get even a few. The possums and foxes and raccoons are probably mostly the inheritors of *Diospyros,* rather than the drivers initially."[12]

Whether American persimmon is regarded as a low-grade, moderate anachronism or as unburdened by such legacy is a question of taste. Does it matter that some legitimate dispersers tend to defecate in plant-unfriendly places? Does it matter that carnivores are not as populous as herbivores and do not have as great a collective appetite for fruits (bear

BLACK SAPOTE. This giant species of the persimmon genus *(Diospyros negro)* is native to Mexico and Central America. Human horticulturalists have helped it persist for ten thousand years. Seed of honey locust (1 cm) and American persimmon for scale at upper left.

being the exception)? However one wishes to make that judgment, persimmon is having little trouble dispersing today. The tree readily invades abandoned fields and is common along fences.[13]

Another kind of persimmon found in Mexico and Central America is undoubtedly an anachronism: its fruit and seeds are of megafaunal scale. This is *Diospyros negro,* commonly called black sapote because of its almost black pulp. (It is no relation to white sapote, of family Rutaceae, or yellow sapote, of family Sapotaceae.) Cultivated varieties of black sapote produce fruits the size of a small grapefruit that contain seeds far bigger than those of *Diospyros virginiana*. Black sapote is grown for human consumption throughout its native range and somewhat beyond. The specimens shown here were grown in Florida. I bought them two blocks from my New York home, for $4.49 per pound. The shelf life of black sapote is very short. These green fruits were firm (and scentless) when I bought them, but two days later they had almost turned to mush—signaling they were ready to eat. The skins were still green, however, and the dark pulp still scentless. Surprisingly, the fruit didn't have much flavor. But there was something about the prunelike color and texture of the pulp that seemed good and healthy in the eating.

Yet another anachronistic persimmon species found in the neotropics is *Diospyros nicaraguensis*. Dan Janzen told me about this fruit in an email. He wrote, "We have a big species of *Diospyros* in the Area de Conservación Guanacaste [in Costa Rica] that dumps tons of fruit on the ground, which just rots today."[14]

I come away with mixed conclusions about the anachronistic status of Western Hemisphere fruits of genus *Diospyros*. Some surely evolved with mastodons and their ilk in mind, others maybe not. Persimmons native to Africa offer helpful comparisons. One wild species native to Ghana, with a fruit diameter of 3 cm and containing six to nine seeds, is taken "incidentally with foliage" by forest elephants, in whose dung the seeds appeared.[15] Yet of twelve *Diospyros* species in a conservation area of Gabon, one absolutely depends on elephants for dispersal. Its fruit can be 8 cm wide and 13 cm long. Chimpanzees are strictly pulp thieves of this big-fruited persimmon; even gorillas rarely swallow a seed. The seeds turn up regularly only in the dung of elephants.[16]

American persimmon reproduces vegetatively with root runners. It can rapidly create a thicket of shrubby growth from just one seed. That talent would have come in handy had persimmon ever been short of dispersal helpers—as may have happened to a plant found in the same habitat as persimmon, the pawpaw. In the previous chapter we looked at pawpaw's capacity to clone. Here we shall examine the anachronistic features of its fruit.

Pawpaws Were Not Made for Pulp Thieves

Pawpaw, *Asimina triloba*, is more anachronistic than American persimmon for three reasons: it has bigger fruits, it has bigger seeds, and the fruits rot soon after ripening. With a shelf life of about three days, pawpaw is sold in roadside stands but not chain supermarkets. The fruit is substantially bigger than that of persimmon. The (overripe) pawpaw shown in the photograph that opens this chapter is about 8 cm long, but some wild varieties can be almost twice that length. Even the smallest pawpaw specimens are too big for foxes and raccoons to take into their mouths whole. Only a bear could process a whole pawpaw with no spillage. Yet the target dispersal agents—those who handle fruits today in the same ways that exerted the greatest selective forces on the pawpaw lineage over its millions of years of evolution—should be expected to take in a fruit whole.[17]

Deer (and presumably elk) will eat neither pawpaw fruit nor the foliage of the tree. There is nothing in the published or Internet literature, however, about whether introduced ungulates—cattle or, more likely, horses—readily eat the sweet fruit and swallow the seeds. What about our slate of midsize carnivores? Mary F. Willson reported in 1993 (using "unpublished data") that pawpaw seeds are defecated "intact and germinable" by raccoons and red foxes *(Vulpes vulpes)*. She cited a 1975 publication in claiming that opossums defecate intact pawpaw seeds too. One should be skeptical of the latter claim, because the purported possum dung examined may have come from raccoons. Misinterpretations of wild-collected dung have appeared in the literature before. If possums do not swallow the seeds of persimmon, then surely they do not swallow the far larger seeds of pawpaw.

I would not want to swallow a pawpaw seed, which is about half again longer and wider than the biggest multivitamin tablet. But it is understandable why animals would if they could. The seeds easily come free of the pulp, but each seed is encased in its own thick and presumably nutritious sack (technically an aril). Whether evolved by "intent" or a lucky byproduct of developmental constraint, the sack is slippery and exceedingly difficult to remove, even by human hands. A quadruped mammal could easily work the sack-encased seeds away from the surrounding pulp once the fruit is in its mouth, but seed spitting would mean losing the calories and nutrients of the aril. To spit or not to spit—if one has the gullet capacity—is a question of balancing losses and gains.

The flavor of pawpaw pulp has been described as a blend of banana, pineapple, and mango, or sometimes, banana, strawberry, and custard. The rich taste is thus similar to that of its most delectable kin to the south, *Annona cherimola,* commonly called cherimoya, which I can obtain from specialty food stores in New York City during the late winter and spring. Cherimoya and pawpaw, both members of family Annonaceae, are as nutritious as they are delicious. In September 1806, during the return trip of the Lewis and Clark expedition, the crew subsisted exclusively on pawpaws for three days while paddling down the lower Missouri River. "Our party appear perfectly contented and tell us they can live very well on poppaws," wrote William Clark in his journal.[18] (Remember, however, that these men had grown accustomed to extreme hardship. During the previous year they had fended off starvation by eating rotten salmon and their own gaunt horses.)

Because the pawpaws I received by mail (from my persimmon source in Indiana) were overripe, I wasn't willing to taste them. Ripe pawpaws are highly aromatic, even cloyingly so, and these pawpaws smelled sweetly rotten. Worse, several had something squirming beneath the skin. Because I don't regard maggot-ridden pawpaws as food, I assumed that our contingent of mesopredators wouldn't either. Wrong. I set out the pawpaws at

New World kin of pawpaw. Dried specimens of tropical species of genus *Annona* from Brazil, Bolivia, Honduras, Jamaica, and the Dominican Republic.

dusk, and they were gone by the next morning. No skin, no seeds—nothing was left. I cannot, however, assume that the seeds were swallowed just because none remained where the fruits had been. Our fox friends undergo a personality shift when night falls. They become quick and wary, possibly because at dusk other creatures begin to visit the compost heap too. Adult raccoons and skunks of any age easily intimidate a fox. So after sundown, if a fox finds something too big to swallow in a single gulp, she or he tends to run off with it to someplace less busy.

I am aware of but haven't become acquainted with our resident raccoons, because unlike the foxes they do not venture forth during the day. Their tracks are unmistakable, however, and I am sometimes roused from sleep when they let out a series of horrifying blasts—not yelps, not howls or screams, but blasts. Somebody took the pawpaw fruits that night, but I cannot say whether the lucky animals were foxes, raccoons, or even skunks.

Whether foxes and raccoons eat around or spit out some or most of the seeds they encounter in pawpaw fruits I cannot ascertain from my observations at the compost pile. I do know that Mrs. Foxie will not swallow the seed of a blue plum, which is smaller than most pawpaw seeds. One day I tossed her four or five blue plums and watched her carefully chew and drop, chew and drop, until all that dropped was the pit. Certainly, the seeds of pawpaw would not be chewed, even if swallowed by native wildlife. Pawpaw seeds (as well as leaves and bark) contain powerful alkaloids harmful to the nervous systems of mammals not equipped to detoxify them. Indeed, an herbal tea made from these parts of a closely related *Annona* species in the West Indies, which is valued for its sedative effects, has been implicated as the cause of a neural ailment with symptoms much like Parkinson's disease.[19] Unlike the seed of an avocado, the seed of pawpaw does not taste bitter. I bit off a seed fragment (the seeds are very tough but not impossible to break), then chewed and spat out the mash. It tasted good in a bland sort of way. That the plant would not have issued a warning that could be sensed by this primate is distressing. Bitter, peppery, or spicy would do. Perhaps I would have been induced to vomit had I eaten and swallowed more, as the seeds have been described as having an emetic effect.[20] Or perhaps my genes have been too long out of the wild to have maintained sensory discrimination. Then, too, maybe we Old World primates have never cohabited with a fruit containing the same toxins that these New World fruits evolved.

Three more signs suggest that pawpaw fruit may be anachronistic, perhaps substantially so. First, the fruits tend to fall within a day of ripening. For the human palate, there is about a three-day window for consumption between under- and overripe (although we can extend this with refrigeration). Mesopredators, as we have seen, find the pawpaw acceptable longer than we do. Maggots are actually desirable, enhancing the protein value of the fruit. But eventually the toxins emitted by molds and bacteria deter even foxes. Especially because pawpaw tends to grow in patches, ideal dispersal partners would be beasts with big appetites, migratory tendencies, and excellent powers of recall. Elephantids excel in all three. The fruit-eating prowess of black bears, however, should not be forgotten.

Another sign of anachronism is that pawpaw is rare in its range, except where associated with indigenous horticulture. Rarity is a sign that the species may have trouble reproducing. A pawpaw patch may be large, but many kilometers might separate one such clone from the next, and successful fertilization may require insect transfer of pollen between clones.[21] Finding legs to help seeds move to a new site is also important. Even more, once a plant goes the route of surrounding its seeds with fleshy pulp, animal helpers become crucial for another reason: to prevent the seeds from rotting along with the pulp. Finally, wild pawpaw trees are usually found in floodplains, yet they grow well if planted in uplands. Floods can, of course, disperse virtually any seed, but only legs can move a big seed uphill.

Based on this compendium of fruit traits and ecological indicators, I propose pawpaw, *Asimina triloba,* as a substantial anachronism, which has survived since the end Pleistocene with the help of a junior varsity team of dispersal agents, human horticulturalists, and its own cloning skills. Its tropical relatives, especially those larger than pawpaw, are also substantially anachronistic. These relatives (all genus *Annona*) include cherimoya, sweetsop, sugar apple, and custard apple. Pending actual experience with this fruit in the wild, my conclusion shall remain tentative, as there is only one published source to back this claim: pawpaw is offered as anachronistic in a single sentence of Janzen and Martin 1982. The scientific study of ecological anachronisms is still in its infancy.

Desert Gourds Were Not Made for Rodents

One fruit not included in Janzen and Martin's 1982 list is desert gourd. For those of us who live in the southwestern states, desert gourd is hard to miss. The fruits are as big and spherical as oranges, and the tangle of vines may convert a large patch of arid landscape into a near monoculture. Desert gourd is a pioneer plant, preferring the sands of floodplains, barren roadsides, overgrazed fields, or the edges of harvester ant clearings. In my rural neighborhood of southern New Mexico, it usually occurs in colonies of several to many individual plants.

Years before coming upon the anachronism idea, I wondered why these gourds go nowhere after ripening. Turning from green to yellow and finally brown, many of the dry or rotted husks can be found in the spring exactly where they were in the fall. They look so much like squashes (and share the same genus) that surely somebody ought to be eating them. Equipped with an awareness of ghosts, I now have an answer: Somebody used to eat the gourds, but those creatures are now extinct.

Of the several species of desert gourd in the United States, the kind that lives around my New Mexico home is *Cucurbita foetidissima.* It ranges as far north as Nebraska, eastward to Missouri, down into Texas, and west to California. It commonly goes by the name *skunk gourd* because crushed leaves and fruit emit a skunky odor. This wild gourd is feared by farmers of summer squashes and pumpkins because its pollen can fertilize the female flowers of the domesticated squash *Cucurbita pepo.* One study reported pollen transfer across nearly a kilometer.[22] When interspecies cross-pollination occurs, the bitterness of the gourd may suffuse the flesh of the squash. A yellow crookneck squash will still look like a yellow crookneck squash, but it may taste a lot like a desert gourd.

A DESERT ANACHRONISM. The desert gourd *(Cucurbita foetidissima)* contains pulp too bitter for humans and well-fed livestock to consume. TOP: Fruit nearing maturation in August. BOTTOM: Fruit rotting and desiccating in January.

Gene flow can go in the other direction, too, and that is probably what has happened in the wild country around my home. The cucurbitacins in the pulp (and, to a lesser extent, the seeds) of desert gourd are among the most intensely bitter substances produced on this planet, and wild gourds contain other toxins as well—saponins, steroids, and tannins.[23] Accidental consumption can give a human intense nausea and diarrhea for days. If a milk cow is pastured on overgrazed land and thus tempted to eat foods she would normally shun, her milk is ruined for human consumption if she munches any part of the desert gourd. Cattle and sheep have died from eating too many gourds.[24] Yet the pulp of gourds around my home that I have chewed (and then spat out, just to be safe) doesn't taste much worse than bad. Bitter, yes, but not much worse than an overripe cucumber (genus *Cucumis* in the same family, Cucurbitaceae). The gourd gene pool in my neighborhood may have been contaminated by proximity to squashes raised on the nearby ranch. More likely the contamination began a thousand years ago, when the upper reaches of the Gila River were settled by native peoples who cultivated squash.

It was, after all, *Cucurbita pepo,* a very close relative of the desert gourd, that various Indian groups bred into summer and winter squashes. *Cucurbita* is entirely an American genus, thought to have originated in Mexico. Especially before pottery was invented, wild gourds would have been valued as containers. Even today, indigenous peoples of Mexico and the American Southwest find medicinal value in the bitter gourds.[25] But they prefer the gourds to reside far away from their cultivated squash fields.

When Paul Martin learned that I was thinking about desert gourd as a Pleistocene anachronism, he sent me a 1993 paper from the *Journal of Ethnobiology* that reported the discovery of *Cucurbita pepo* seeds in a Florida bog. The seeds were found in association with the skull of a mastodon and also with intestinal contents that are presumably mastodon.[26] The authors suggested that proboscideans had been dispersal mutualists for this genus, and they cited Janzen and Martin 1982 as support.

The pulp of *C. pepo* may have evolved as mastodon and other megafaunal bait. In Africa, seeds of diverse members of family Cucurbitaceae are prevalent in elephant dung.[27] The undomesticated, wild variety of *C. pepo* is bitter to the human palate. The bitterness is there to keep undesirable characters—vertebrates or insects—from stealing the pulp or raiding the seeds, but is it intended to deter everybody? Surely not. Apparently it did not deter the Florida mastodon.

The same should hold for *C. foetidissima.* We know that introduced livestock (proxy Pleistocene megafauna) will consume desert gourds in moderation, especially if the animals are desperate, and one report suggests that burros are more inclined than the rest to regard gourds as food.[28] To support a conclusion that desert gourds evolved with extinct megafauna as mutualists, it would be helpful to point to fruits in the Old World whose bitterness only the megafauna can tolerate. The fruit of *Trewia nudiflora* of lowland Nepal may provide such comparison. As described in Chapter 3, rhinoceros is the tree's chief and nearly exclusive native disperser. Trewia fruit is reportedly "extremely bitter to our taste."[29] Wild rhinos eat them not as a last resort but as a preferred food. Taxonomically closer to desert gourd are the wild African cucumbers that forest elephants reportedly take.[30]

If they are not evolved to attract megafauna, then how else does one explain the apparently overbuilt character of *Cucurbita foetidissima*? Desert gourds contain a mass of pulp suggestive of a function. If not food for a mammalian mutualist, then what?

Gary Paul Nabhan, ecologist and ethnobotanist of the American Southwest and northern Mexico, offered a fine but shortsighted just-so story. "Desert gourds may linger in the same place for months," he wrote in 1987, "shrinking in size and hollowing as the bitter flesh dries into a stringy mass around the seeds. Then, when a rain brings a torrent to their floodplain abode, they suddenly find buoyancy and ride the crest of a flashflood."[31] All this is true, but it is nothing out of the ordinary for a fruit. Even fresh gourds full of moisture are excellent floaters. I've had one floating in a bucket of water for almost two weeks now.

Nabhan continues the story of the flood-swept gourds: "As they crash into boulders, hackberry trunks and arroyo banks, the gourds' brittle shells are punctured, pounded, and pulverized into irregular pieces. Seeds are scattered by the waters and scarified as the flood bedload scours their seed coats. They may end up in any place the water has washed over, often under several inches of rich, flood-carried silt and detritus."

True again, but we are still no closer to solving the mystery of the evolution of desert gourd. Any fruit will splinter in a flash flood, and it shouldn't need to invest in a lot of pulp to ensure a favorable outcome in such circumstances. A paper published after Nabhan told this imaginative story reports that the seeds of desert gourds do not require scarification in order to germinate.[32] A flood may liberate and scatter them, but gourd seeds need not crash into boulders. Floods may be a primary means of dispersal today, but did the fruits evolve with floods as the aim? That a fruit is watertight and buoyant may not owe to selection by flowing water;

rather, the fruit is incidentally adapted—preadapted—for this dispersal path. Impermeability is, after all, helpful in a desert environment and wherever the spores of molds may be carried on a breeze. Oil is in the seed not for buoyancy but to feed the sprouting plant.

Two observations support the contention that the fruit of desert gourd was not shaped by flowing water. First, desert gourd is not the only fruit afloat in my bucket. The fibrous sphere of osage orange, along with the stiff pods of honey locust, mesquite, and Kentucky coffee tree, has also been floating there for twelve days. Just now I was inspired to extend this experiment. I went to the kitchen and collected a sample of every fruit I could find. One at a time, I dropped them into a large bowl of water. The best floaters were lime, banana, avocado, and apple. A good part of each remained exposed above the water line. Tomato hovered just below the surface, and pear grazed the bottom. Kiwi was the only fruit that sank outright. Of the two gourd relatives I happen to have on hand, the acorn squash bobbed beautifully, while the butternut squash floated only at the end that contains oil-rich seeds amid pockets of air. Conclusion: buoyant fruit is not unusual. To observe that a fruit can be carried by moving water is no more helpful than to observe that gravity draws a fruit from canopy to ground. The ability to float and the ability to fall do not owe to natural selection. Floods may fortuitously help a fruit lineage survive the loss of its animal mutualists, but a talent for going with the flow was not the aim of evolution, at least not in these cases.[33]

A second observation that argues against the flood dispersal story turns on where desert gourds can be found in the wild. Yes, they are often in the sandy stretches where the river floods (or where the river used to flood before it severely downcut the riverbed), and I have seen whole and partial gourds amid the debris left by high waters. But I have also found gourds (though rarely) well into the uplands. This summer I came upon two isolated gourd patches in an otherwise grassy meadow, about two hundred meters above the river and three kilometers inland. It is possible, just possible, that these plants are remnants of an upland population that thrived when the megafauna were here. Or maybe the pre-Columbian Indians intentionally or inadvertently planted them at the upland site. A final possibility is that the gourds were deposited in the dung of proxy Pleistocene megafauna during the time, not long ago, when rampant overstocking of public lands compelled livestock to eat poor-quality foods.

Another issue: If a fruit is dispersed primarily by floods, how do progeny return upstream? Where Nabhan and I both get stuck in our stories are roadsides. Roadsides are ideal for these pioneer plants, but how do they get there? The road into our village was built and paved in the 1960s.

Today it is rimmed with many patches of desert gourd, though far fewer than the number of bare spots that would readily accept them.

Nabhan was not the first to be fooled into thinking that flotation is an adaptation rather than a by-product of evolution. In a 1982 paper, Dan Janzen criticized a colleague who had "overlooked the role of Pleistocene megafauna to such a degree that he viewed *Crescentia* as water-dispersed despite its sweet fleshy interior."[34] The tough-rinded, spherical fruit of the neotropical *Crescentia* tree is indeed an excellent analog for desert gourd. The seeds of both are exquisitely insulated from insect attack. But a suit of armor can suffocate too. Janzen found that for *Crescentia,* "If the fruit is never broken open, the seeds die of desiccation without germinating, if the habitat is dry. If the site is very moist, unopened fruits rot and the fruit pulp–seed mass inside ferments . . . which again kills the seeds."

We are so accustomed to ignoring the past that our view of ecological interactions in the present is seriously askew. Gary Paul Nabhan may have been fooled by desert gourd initially, but he wasn't fooled for long. Soon after publishing his gourd story, Nabhan got onto the track of ecological anachronisms (thanks to his association with Paul Martin) and has been there ever since. A talk he presented at a 1987 ethnobiology conference is subtitled "Did Mesoamericans Disperse and Select Anachronistic Fruits?"[35] Nabhan's ideas of how humans have helped anachronistic plants persist are dealt with in the final chapter. For now, it is important simply to note that, like Dan Janzen before him, Nabhan let go of a good tale in order to begin telling a more truthful—and more exciting—story.

My own understanding of the adaptive features of desert gourd has come a long way since I first puzzled over the riddle of its rotting fruit. On balance, a plant that grows well in dry uplands (as desert gourds do) yet is mostly found in floodplains should be considered a prime suspect for anachronistic status. What about other survival skills? What about vegetative forms of reproduction? And what about second-string dispersal agents? Nabhan has not pursued these questions. Nor has Janzen or Martin.

Of the twenty-some relevant abstracts I have collected on *C. foetidissima,* one ascribes clonal properties to this species: "Within patches, plants of the same sex type usually shared identical five-locus genotypes, suggesting that clonal propagation predominates."[36] As to backup dispersers, by far the most intriguing is a squirrel-size rodent. Having seen only the results of the interaction, I cannot be sure of the rodent's identity. Nevertheless, I surmise that the backup disperser is either a rock squirrel *(Spermophilus variegatus)* or, more likely, a pocket gopher *(Thomomys bottae)*. Evidence of the interaction was a collection of sev-

Rodent reaping the benefit of the megafaunal extinction. If North America's megafauna were still alive, the rodent that dug this burrow would have had to work a lot faster to hoard these desert gourds. It took two weeks for the rodent to collect the gourds and another two to store them within the burrow.

ered gourds perched on a great mound of sand at the entrance of a burrow 15 cm in diameter. I made this discovery within a gourd patch that covers half a basketball court of sandy river terrace near my home. On September 25 I counted 14 freshly ripened fruits at the entrance to the burrow. I visited the burrow again four days later, and this time counted 84 gourds. There was no sign that seeds had been plundered from any of the fruits at the burrow entrance; all were intact spheres. By October 5 the gourds at the entrance numbered 76. Five days later the tally dropped back to 14. October 15 yielded 8. By the next visit, October 31, the last had vanished.

There was one other burrow, four meters away, where a similar spectacle (though of lesser degree) had been taking place. Virtually all mature gourds within five or six meters of each burrow had been harvested, but none had been taken from points beyond. Perhaps this was a bad year for rodents. There is no cover, save the low vines, to protect them from hawks or owls when they venture beyond their burrows. Clearly, rodents had been gathering the gourds, but where was the final stash? There was only one place to look: inside a burrow.

During the October 31 visit, I removed the plug of sand at the entrance, reopening the tunnel. I reached inside, following the tunnel's arc to the left. Where the tunnel straightened, I came upon a smooth, curved object. I removed it, then another, and another, until I could reach no farther. The five retrieved gourds were in good shape, though covered with moist sand. They were the only objects I encountered in the portion of the tunnel within reach.

Photography finished, the gourds were returned to the pantry and the sand plug restored. The mystery had been solved—or at least part of it. I surmise that the roadside patches where vines die back in the fall to reveal few or no gourds are in places either where male flowers prevail or where burrowing rodents could find safe homes nearby and possibly sandy earth for storing gourds underground. Confirmation of these hypotheses will have to wait. For now, the quest appears endless. Finding an answer to one question inevitably turns up two or three more. For example, I would love to learn more about another kind of desert gourd, *Apodanthera undulata*. It belongs in the same plant family as *C. foetidissima:* the family Cucurbitaceae. Paul Martin introduced me to it while we were botanizing along the Arizona-Mexico border. And of course, we both began to conjure ghosts in its presence.

I will leave to another enthusiast the task of evaluating *A. undulata* as a candidate anachronism. *C. foetidissima* is the only desert gourd I am willing to pass judgment on. I propose for it the grade of substantial anachronism, perhaps even extreme anachronism. The only animals possibly willing to swallow the pulp (the record is unclear) are nonnative horses or burros—proxy Pleistocene megafauna—and it appears these animals must first be moved by hunger in fenced, plundered pastures. Rodents do hoard some fruits in underground larders, but that concentrates rather than disperses the crop. The great majority of gourds remain unharvested. Some are raided for seeds during the coming months, while the rest rot or desiccate where they lie. Because rodents do not consume the bitter pulp, the fruit of desert gourd is unarguably overbuilt for them. It is thus reasonable to conclude that desert gourd was not made for rodents. Floods do carry the fruits, but buoyancy is not an adaptation *for* flood dispersal. Then, too, this desert gourd has an extremely patchy distribution.

Someone needs to study *C. foetidissima* (and *A. undulata*) as thoroughly as *Hymenaea courbaril*, another rodent-hoarded fruit, has been studied in Costa Rica. Locally known as guapinol, it has a thick woody pod that is filled with dry sugary pulp of the same pistachio green hue as honey locust's. (Guapinol is one of the seven pods portrayed as "compelling anachronisms" in the photograph on page 82; it is the fat pod nearest my hand.) Guapinol has been pronounced an anachronism, saved by its invulnerability to bruchid beetle attack and by the hoarding services of a neotropical rodent, the agouti.

The 1982 Janzen and Martin paper on neotropical anachronisms stimulated a two-year field study on the agouti-guapinol relationship. Winnie Hallwachs concludes: "The present interaction between guapinol and its seed predators and dispersers is serendipitous rather than coevolved: guapinol fruit traits probably have changed little since they evolved in the Oligocene [thirty million years ago] among a species-rich fauna of large herbivorous dispersal agents."[37] She continues, "Guapinol survived as a widespread and common tree largely because agoutis (who may have been minor members of the guapinol's pre-Pleistocene disperser coterie) continued to disperse its seeds after its megafauna dispersal assemblage died out. . . . The ancient tree *Hymenaea* did not evolve for agouti dispersal, but became dependent on agoutis when the megafauna went extinct."

Hallwachs ends her report with suggestions for more study. Among the fruits she offers as candidate anachronisms are two that were mentioned in previous chapters of this book: *Gustavia superba* and various species of *Persea*. *Gustavia,* you may recall, is the neotropical fruit that turned Henry Howe from a formidable critic of the anachronism idea into a partial supporter—but only for a pared-down list of candidates. *Persea* is the avocado genus.

Thus far I have told stories of four anachronistic fruits in temperate North America: desert gourd, pawpaw, persimmon, and (back in Chapter 2) honey locust. Each story drew upon a combination of published information and my own observations and crude experiments to assess the physical, chemical, and ecological traits indicative of anachronism. Why, you may wonder, doesn't she do the obvious? Inferences from such data may be helpful, but how about offering the suspected anachronisms to real live Pleistocene megafauna—especially the megafauna of the grandest proportions? Why not try out these fruits on zoo elephants? As we shall see in Chapter 8, there is a world of difference in the food preferences of wild and captive creatures.

DREAMING OF DINOSAURS. Ginkgo's seed dispersal strategy may be adapted for dinosaurs. The pale orange fruit smells like vomit when ripe; the single seed is protected in a nutlike, but thin, shell. Humans rescued this tree from the brink of extinction. A medicinal capsule of ginkgo leaf extract is shown for scale.

6

Extreme Anachronisms

THE DISCUSSION OF WILD PERSIMMON IN THE PREVIOUS CHAPTER may not have convinced you that anachronisms exist. At most, the persimmon is moderately out of step with time; foxes, raccoons, and bears reliably disperse its seeds. The anachronism argument is stronger for pawpaw, as the same locally based animal associates who disperse persimmon over the course of several months must rise to the challenge of consuming the bulk of the pawpaw crop in just a few weeks. The smaller among them eat around more seeds than they swallow. The desert gourd presents an even stronger case for anachronism. Possibly nobody except introduced livestock is defecating gourd seeds, and then perhaps only in harsh, unsupplemented range.

It is thus time to call in the extreme examples. If an indisputable case can be made for the presence of ecological anachronisms in the United States, it will be carried by two enormous fruits that absolutely no native species regards as food. Indeed, absent the anachronism perspective, one would have to judge these fruits evolutionary enigmas. One fruit is a dense, fibrous sphere containing hundreds of seeds. The other is a nearly impenetrable pod that encases a half dozen big seeds in a bed of sugary pulp. Both these strange fruits have compensatory life history traits that can explain their survival without animal partners for thirteen thousand years.

Thirteen millennia is nothing next to tens of millions of years, however. Osage orange and Kentucky coffee tree may remember the mastodons, but ginkgo—the ultimate extreme anachronism—surely dreams of dinosaurs. Ginkgo is not currently native to North America (growing wild on just a few mountains in a remote part of China), but it used to be native to temperate regions throughout the Northern Hemisphere. The only remaining species of a once magnificent family of trees that thrived in the Jurassic and Cretaceous, ginkgo was rescued from oblivion by the animal better known for extinguishing biodiversity: us. It was able to hold on during the interim because of clonal capacities that verge on the miraculous.

The ginkgos, osage oranges, and Kentucky coffee trees of the world are the plants that will secure the validity of the theory, convincing even diehard skeptics that ecological anachronisms must be acknowledged. Only a wealth of examples more on the line of pawpaw and desert gourd will attest to the theory's importance, however. Far more examples of all grades, from moderate and substantial to extreme, exist in the tropics than in the temperate zone. But the temperate zone of North America is where I live.

The Enigmatic Osage Orange

"Take as many as you'd like. The horses won't touch them, so they're just a nuisance." Becky Campbell, my neighbor in New Mexico, is speaking of the fruits that fall in October from the two female osage orange trees that border her irrigated pasture, along with three males. Unlike honey locust, whose pods remain aloft for months after ripening, the fruits of osage orange fall pretty much all at once. And there they lie, taking an awfully long time to rot. If osage orange (*Maclura pomifera,* henceforth maclura) is waiting for ghosts, the missing partners need not have had proboscidean trunks.

"Unpalatable to man or beast." "Coarse pulp and sticky latex render the fruit inedible." "Little wildlife value except to provide nesting and hiding places."[1] These are the sorts of things one reads about maclura. This, too: "The general assumption is that osage orange fruits are poisonous. There is little evidence to support this notion. Pullar (1939) reported that the fruits were toxic to sheep. Death of horses and cattle attributed to osage orange are more likely the result of choking on the sticky fruits rather than to actual toxicity."[2]

Given these descriptions, I was reluctant to taste maclura. But curiosity proved stronger than prudence, especially because I couldn't find any mention of what the fruit actually tastes like. Exactly why is it "inedible"? I began by slicing one of the grapefruit-size, glow-green balls, then dabbing onto my lip the white latex oozing profusely from the cut. No problem, so I tried the latex on my tongue. Still no problem. I took a chunk of fibrous pulp and chewed. The fruit tasted surprisingly good, but more like air freshener than food. Strange though it may sound, the taste was delightfully clean, with perhaps a hint of cucumber. Several publications depict maclura fruit as bitter, but that was not my experience.

An extreme anachronism. Osage orange, *Maclura pomifera,* is one of North America's most anachronistic fruits. Freshly sliced fruit oozes a white latex, which has been wiped clean from the slice on the left. Honey locust seed (1 cm) for scale.

The latex of maclura, however, can be messy. My fingers became very sticky in a way that water alone could not dissolve. Not even dishwashing soap would do the job. Only a strong laundry detergent seemed to work. Like honey, though, the latex melts in the mouth, thanks to the magical powers of saliva.

Paul Martin gave me the courage to start tasting wild plants. While roadside botanizing with him in South Dakota and again in Mexico, I was intrigued by how often he chewed a leaf to confirm an identification. So I asked for instruction in the art. Later he tested me. Drawing my attention to a tall herbaceous plant that was neither flowering nor fruiting, he asked, "What's this?"

I plucked a leaf and chewed. "I would guess a composite [daisy family]. It tastes like turpentine."

"Right you are."

I also learned from Paul that it's a good idea to spit out what one has chewed. So now I taste "inedible" fruits in the same way one is supposed to sample wine when there is a lot of wine to sample. In the case of maclura I was especially wary because the fruit is sold in green markets as a kitchen aid to repel cockroaches. If a burro won't eat it and a cockroach won't go near it, one should indeed be cautious.

Two weeks after my first taste of maclura, the tops of the fallen fruits in the pasture had turned a bold yellow from frosts. When first fallen, maclura is bright green and hard as a raw potato. But now, although still firm and with no sign of rot, they had become slightly spongy. Another taste test was in order. Surprise! No latex oozed from the wound this time. The cut fruit had the same lovely smell as before, although not as intense as when it had just dropped. Because the fibrous pulp was drier now, my primate fingers were able to separate the radially arranged compartments with little effort. Every third or fourth compartment had a plump seed at its base. All seeds were embedded within the fruit about 2 cm beneath the surface, and there was just this one layer of seeds. Because of the depth, a seed predator would have to do a lot of gnawing to extract even one small reward. The design of an osage orange is thus very similar to that of citrus fruits. Slice a grapefruit through its equator (especially the breeds that are still very seedy) and you will find that the seeds occupy a single layer near the core.

For this test, I counted the seeds before popping a chunk into my mouth. The ripe fruit still tasted like air freshener, but now the flavor was muted and carried a hint of citrus. There was absolutely no bitter taste or aftertaste. I chewed that chunk until it had softened to the point that swallowing seemed the right thing to do. During that time, I was barely aware of any seeds. Examining the mass, I found three seeds intact, with no seed fragments. I took a second chunk and determined to chew it more thoroughly this time. Late in the process, my molars began to crunch seeds, which are slightly smaller than the seeds of commercial oranges and about the same resistance. The seed coat is no deterrent. A crunched seed sharpens the taste, but the result is not repulsive. Perhaps it is repulsive to wild creatures. The seed contains lectins that glue red blood cells into useless masses. Lectins are potent toxins against mammals, birds, and even some insects.[3]

Again, I spat out the mash. This time there were three crunched seeds. Only one survived the ordeal. My hypothesis: Maclura is hoping to attract an alimentary canal substantially bigger than mine, who could take the whole fruit into its mouth all at once and would chew the pulp less than thoroughly. That animal (or, more likely, animals) must have had better detox systems than mine. Like the fruit of Costa Rica's *Crescentia,* a crop of fallen maclura would take several weeks to ripen to the point that animals would begin to show an interest. And like *Crescentia,* the crop could wait a long time to be harvested. There simply are no pulp thieves, and seed predators wait for the stickiness to subside.

Rodents are fond of maclura seeds, but that doesn't mean the fruit was designed with rodents in mind. Our resident mouse gathered any bare seeds it was offered, but I had to do the extracting. Whether they find the latex physically or physiologically repulsive, rodents seem to prefer attending to other autumn tasks first.[4]

In early January, while visiting friends, I chanced upon an osage orange tree in the botanical garden at Fort Worth, Texas. Two green fruits still hung from a branch, and one was on the ground. The rest had been shredded. Somebody—presumably a squirrel—had dragged all the balls to the base of the tree. There the animal had sat, tearing apart fruits in quest of seeds. The pulp had oxidized to a muddy maroon, but there was no overt sign of mold or rot among the fragments. The discarded pieces of pulp now made a halo around the trunk, like a blanket encircling the base of a Christmas tree. Where it touched the trunk, the pile was almost knee-deep, and it contained many empty seed cases. My hunch is that squirrels don't bother to store the seeds in a larder or to bury them in scattered caches. There is no safer store than the intact fruit itself. Squirrels probably gather a fruit from beneath the canopy only when hungry, drag it back to the safety of the trunk, and then shred the pulp.

Published reports are unanimous in claiming that humans do not eat the fruit of maclura—not boiled, roasted, or otherwise processed. The reports are nearly unanimous in stating that no animal—native or introduced—will eat the fruit either. All that pulp must be there for a reason. Perhaps maclura is playing an evolutionary game of attraction/repulsion. On the one hand, energy-rich pulp in such bulk should attract herds of big animals who can be depended upon to deposit ingested seeds well beyond the shade of the parent tree. On the other hand, maclura wants to repel everybody else. Despite some toxins, the seeds are attractive to seed-eating rodents. The embryo and its sack lunch of endosperm are encased in a shell that is easy for seed predators to breach, if only they can get to the seed. Maclura therefore deeply embeds its potential progeny within a dense, sticky pulp that repels or at least frustrates rodents who might otherwise rip apart the fruits before the mastodons arrive. Maybe that strategy is also intended (or was primarily intended) to prevent bruchid beetles from using the seeds as a nursery. Hatching from an egg deposited on the fruit surface, a tiny larva would have a long and gummy course to navigate in quest of a seed.

Whatever the actual adaptive story, the fruit of maclura is conspicuously overbuilt if repelling rodents and insects is the sole intent. I can attest to the unworthiness of field mice as understudies for the mastodons, and as I saw in Texas, gravity alone would be more helpful than a squirrel. One can thus conclude that maclura carries the traits of a megafaunal fruit. What about ecological indicators? Are there contextual clues that suggest maclura may be anachronistic, perhaps extremely so?

The native range of maclura prior to the arrival of Europeans is thought to have been the Red River region where Arkansas, Texas, and Oklahoma join. This is a very restricted and spotty range, since fossils indicate that during interglacial episodes of the Pleistocene epoch the tree lived as far north as southern Ontario. Maclura thus reveals the dangers a plant faces when its long-distance dispersers go extinct. Elephants in Africa not uncommonly disperse a seed fifty kilometers from the parent tree.[5] Long-distance dispersal may be vital over the long term to ensure recolonization of sites after catastrophic events, or to facilitate climate-induced migration across habitat barriers.

Native Americans placed tremendous value on osage orange trees because the wood is unsurpassed by any wood in the world for making bows. In 1804 Meriwether Lewis dispatched a letter to President Thomas Jefferson from St. Louis, just before commencing the Corps of Discovery expedition into the unexplored West. He wrote, "So much do the savages esteem the wood of this tree for the purpose of making their bows, that they travel many hundred miles in quest of it."[6]

Lewis's statement, coupled with our knowledge of the restricted range of maclura, argues against a history of intentional planting by Native Americans. Maclura easily becomes naturalized outside its native range—meaning that wherever it is planted, it not only thrives but spreads. So well does this tree grow in Indiana, for example, that the fruit has been nicknamed "Indiana brains." Maclura now flourishes wherever it was planted for hedgerows before the invention of barbed wire. Today it grows beyond the confines of hedges as far north as Colorado and throughout almost all the eastern states. Maclura has not only adult humans to thank for this turnaround in fitness but probably also children. What child can pass by a carpet of brilliant green softballs and not be moved to give at least one a toss?

In addition to range shrinkage, another indicator of post-Pleistocene troubles is that the genus *Maclura*, which today consists of only one species, had entailed several species in North America.[7] Indicative, too, is that in its native range, before it was propagated by European colonists, maclura was found mostly in floodplains, possibly exclusively in the rich bottomlands of the Bois d'Arc Creek drainage of the Red River. Today, however, this sun-loving tree readily invades pastures and fallow fields. "The osage orange has found the valleys of the Ohio and of other northern rivers quite to its liking; it is now common in many places where it was not planted, particularly along river and creek banks where running water has served to distribute its fruit."[8] Yes, running water can distribute the fruit of maclura, but as we saw in the previous chapter, buoyancy is a by-product—not the target—of evolution. As I write, the maclura fruit in the bucket has been floating for sixteen days—but then, so have the fruits of honey locust, mesquite, and desert gourd.

What special skills does maclura possess that could have sustained the lineage after its dispersal mutualists disappeared? The tree can be long-lived; the wood is extremely resistant to rot. The website of the U.S. Department of Agriculture reports that "osage orange heartwood is the most decay-resistant of all North American timbers and is immune to termites." Decay resistance, coupled with superior hardness, made maclura an extremely valuable wood during the nineteenth century, especially for railroad ties and the hubs and rims of horse-drawn carriages.[9]

Another useful skill for an extreme anachronism is the ability to reproduce vegetatively. Maclura is adept at resprouting multiple stems after the main stem has been injured. Like pawpaw, it also produces shoots from lateral roots that may venture many meters from the elder stem. Andrew Schnabel and colleagues found, however, that the stems within clumps of maclura growing in abandoned pasture near an old hedgerow in Kansas were not clonemates but sibs. Honey locust clumps scattered in the same field were also analyzed for genetic relationship. The same pattern held. The authors concluded that "this pattern of spatial structure for both species results from extremely local seed dispersal and establishment of individuals from the same multiseeded fruit." As explanation, they explained it was common for "dispersers to sit directly at the base of trees while extracting seeds from fruits."[10] (Sounds like seed predators to me.)

Maclura has another trick that surely bolstered its persistence. It can produce seeds apomictically.[11] This means that a female tree need not receive pollen from a male in order to produce fruit with viable seeds. Each seed will be her clone, and it will look just like a standard, sexually produced seed embedded in a standard fruit. In this way, maclura has followed the evolutionary path of dandelion. The fluffy parachutes that dandelion launches on the wind dangle seeds that are almost exclusively the product of apomictic reproduction.[12] As we have seen with pawpaw (which bears both sexes in the same flower but cannot self-pollinate), one of the dangers for a species ebbing toward extreme rarity is that, at some point, pollen becomes scarce. If there are no male anthers upwind, the ova will not be fertilized. But species that develop a capacity to self-pollinate or that can reproduce apomictically go ahead and produce seeds and fruits anyway. Maclura is thus well positioned for survival in a world of ghosts.

All of these factors—fruit structure and chemistry, ecological indicators, and compensatory life history traits—lead me to regard maclura as anachronistic in the extreme. Are there ways to test this contention? One way would be to find an analog in the Old World that matches maclura in form, texture, and reward. Is this analog attractive to Old World megafauna? Is its pulp repulsive to lesser creatures like us?

The genus *Maclura* is native only to temperate North America. Whether it can hybridize with the nearly identical genus *Cudrania,* which is native to China, Korea, and Japan, is a matter of dispute.[13] Both genera hail from the same, largely tropical family as mulberry, figs, and rubber (family Moraceae). Such fruits are compound, containing many seeds embedded in pulp, and each seed is provided its own little home, walled off from contact with its fruit-mates.

Mulberries are small, juicy, and dispersed by birds. The fig strategy is to appeal to primates and bats while still on the branch and to just about everybody after fruit fall. Maclura doesn't resemble either of these fruits except in its developmental pathway. It does, however, strikingly resemble the breadfruit and jackfruit of Asia—both also members of family Moraceae, genus *Artocarpus.* Many of the fifty species of this genus produce fruits that are considered unfit for human consumption, and some bear fruit as small as a cherry. The two widely cultivated species, breadfruit and jackfruit, have been domesticated for so long that it is difficult to determine whether the ancestral stock was enormous like the descendants. During the European invasion of the American West, had the white settlers been familiar with a wider range of tropical fruits, maclura might now commonly be called osage breadfruit.

A friend of mine from Kansas, who in 1968 was assigned to the Asian tropics for training as a Peace Corps volunteer, wrote of his encounter with breadfruit. "When I saw my first fresh-picked breadfruit, it reminded me of a hedge apple; the same tough, green, bumpy-textured skin, except three or four times larger; same on the inside too—white, rubbery, and stringy (except you could cook and eat the insides of a breadfruit). It wasn't until years later that I looked both trees up in the encyclopedia and discovered that they are first cousins in the mulberry family." John Brewer reflects on his response to the exotic cuisine: "Amidst the assortment of stronger flavors and brighter colors, breadfruit was a bland but dependable presence. Little did I realize that over the months and years of my life in Pohnpei, it would become my favorite food not only in Micronesia but in the world."[14]

Breadfruit—thought to have originated in Indonesia, Java, or New Guinea—is globose or oblong and can be the size of a soccer ball. Many of the breadfruit cultivars have been selected for seedlessness and are propagated from root cuttings. Other varieties bear seeds as large as a chestnut and just about as savory when roasted. Some variants are cultivated only for their seed, with the pulp reserved for livestock. Breadfruit is harvested in a mature but underripe state and always cooked.

Jackfruit, inferred to have originated in Malaysia or India, looks like a breadfruit of uncommon size. The pull of gravity has made it pendulous, and it can grow to nearly a meter in length. Like breadfruit, an underripe jackfruit can be consumed if cooked, but jackfruit pulp is also eaten raw when fully ripened. Its seeds are valued as food too. Breeding practices have created a range of jackfruit flesh, from moist and pulpy to dry and fibrous. Like maclura, jackfruit and breadfruit both contain milky latex at some stages in their development and ripen to a greenish yellow, with a slight spongy give.

If these are not megafaunal fruits, I don't know what is. Unfortunately, the literature is silent on visitors to *Artocarpus* trees in the wild. Asian fruit-elephant mutualisms of any kind have received little attention.[15] Fortunately, Africa also has an indigenous analog of maclura, and much work has been done there chronicling the fruits dispersed by forest elephants. *Treculia africana* is the "African breadfruit." Like its sister genus *Artocarpus, Treculia* is cultivated. Africans regard the pulp as poisonous to themselves, so it is discarded. Sometimes the pulp is fed to livestock, but injury may result. Only the seeds are valued as human food, but they must first be cooked to denature the toxins.[16] Despite its harmful components, *Treculia* pulp is favored by forest elephants in Gabon. (Physiological and behavioral tricks elephants possess for dealing with the toxins are discussed in Chapter 8.) *Treculia africana* is the biggest of all elephant fruits in Africa, with a diameter sometimes exceeding 25 cm.[17]

The week before this book manuscript left my hands for production, Paul Martin alerted me to a just-published paper that beautifully confirms maclura's anachronistic status.[18] Frank F. Schambach at the Arkansas Archeological Survey wrote in passing that horses "relish the fruit of osage orange." That statement did not compute with everything else I had read or with the experience of my horse-ranching neighbor. What was Schambach's evidence?

It turns out that horses pastured on a ranch in maclura's traditional habitat near Texarkana, Texas (right along the Arkansas border), were very happy to eat any osage orange fruits Schambach offered them—even the freshly fallen green fruits. Schambach, an archeologist, was forced to undertake this casual experiment because he, too, was frustrated by the anecdotal nature of published evidence on the matter. The outcome would affect his piecing together the puzzle of Indian trade routes and the distances over which bows carved from maclura wood were traded.

"Botanists just aren't paying attention to this," he told me. When I read to him the ancient reports I had found about livestock "choking to death" on the fruits, he scoffed. "That may be true for cattle, as cattle won't eat them, but not horses."

Why won't cattle eat maclura? Schambach didn't have a ready answer, but I proposed that their lack of upper incisors may be the constraint. Cattle could not bite the fruit into chunks small enough to grind with their molars. So any cow tempted to eat maclura fruit would have to swallow the ball whole, with choking being one possible outcome.

No, he hadn't examined horse feces to look for viable seeds. And no, he hadn't tasted maclura and didn't know of anybody who had. But he was interested to hear my personal account. We talked, too, about the possibility that the shrunken range of osage orange might owe in part to Native Americans' regarding maclura saplings as valuable bow material, coupled with the absence of any evidence that they might have intentionally planted maclura for future use.

With far more confidence than undergirds my judgment of persimmon as a moderate anachronism and pawpaw and desert gourd as substantial anachronisms, I propose the fruit of osage orange, *Maclura pomifera,* as an anachronism of an extreme degree. It exemplifies the megafaunal dispersal syndrome. It has analogs in Asia and in Africa that are close relatives and that are in fact consumed by megafauna. No animal native to North America regards the pulp of maclura as food, and introduced horses may or may not. The fruit simply rots where it falls, the seeds undispersed—save for the rare assistance of seed predators, whose hoarding services (if any) probably fail to move seeds beyond the shade of the parent tree. In addition, the native range of maclura seems unduly restricted in light of the tree's geographic success during its partnership with humans the past two centuries. Root suckering, apomixis, and exceptional resistance to wood rot and termites are compensatory life history traits that could have held extinction at bay for thirteen thousand years.

Which story, then, demands the stronger argument? Shall it be taken that the fruit of *Maclura pomifera* is well adapted to present conditions, because a causal explanation rooted in the past cannot be decisively tested? Should we sidestep the issue altogether by invoking phylogenetic inertia as the reason the fruit is as it is, without attempting to explain its strangeness? Or would it be prudent to conclude that the fruit evolved to entice a cadre of large beasts who are now ghosts?

The Strange Case of Kentucky Coffee Tree

"If in your walks through the woods, you chance to come upon a Kentucky coffee tree, count yourself fortunate, for it is one of the rarest of our forest trees; not that its range is limited, for it is found from southern Canada to Tennessee and from western New York to Minnesota and on down to Arkansas, but that it is nowhere abundant."[19]

This and other clues to the anachronistic status of the fruit of Kentucky coffee tree, *Gymnocladus dioicus*, can be gleaned from reports published long before Dan Janzen and Paul Martin initiated the concept. Another ecological indicator of anachronism is that the tree is mostly found in floodplains, even though it grows well in uplands too—thus suggesting reliance on flowing water for dispersal. A 1935 report offers this intriguing analysis:

> The fruit has no special means of dispersal; the wind, even if strong, cannot carry it far, for the pods are too thick and too heavy. Rodents are not known to store them away for food. No doubt, at occasional flood-times, the creeks and rivers do carry the pods some distance downstream, but, in whatever way it is done, the natural distribution of the fruit is scanty compared with what squirrels do for the walnut, birds for the red cedar, and the wind for maple. And yet, though nowhere common, it is found over a very wide range in eastern America. To account for this we must take into consideration its unusual vitality and the many thousands of years it has been one of our forest trees.[20]

Tough as old leather, the pod cannot be chewed by the likes of me. Even our resident mouse, who harvests seeds from any mesquite or honey locust pods left about the porch, will not bother to gnaw through a gymnocladus pod. The seeds are protected by a coat even tougher than that of honey locust, and they are the largest of any seeds borne by anachronistic fruits in the United States. A gullet larger than mine would be required to swallow them. What is the reward?

Like honey locust, gymnocladus nestles its seeds in a bed of green pulp. To me both kinds of pulp taste the same: exceptionally sweet, with a dash of bitter. Unlike honey locust, however, the pods of gymnocladus are not fed to livestock. They are regarded as poisonous. The pulp contains saponins and alkaloids. In China the pulp of *Gymnocladus chinensis* is used for washing clothes because the saponins become soapy in water. Most lengthy descriptions of gymnocladus include the statement that one woman in Kentucky did, in fact, die from eating pulp, but the story is oft-

THE MOST ANACHRONISTIC LEGUME IN NORTH AMERICA. The ripe pod of Kentucky coffee tree, *Gymnocladus dioicus,* is toughened by resins, and the seeds are invulnerable to insect attack. The green pulp is sweet but reputedly poisonous to humans. Early colonists in Kentucky roasted and ground the seeds to make a coffeelike brew. The ground "coffee" shown here is courtesy of Carl Mehling.

repeated hearsay. As with fruits in general, the deadliest poisons may be concentrated in the seeds.

"Keep children away from the pods of Kentucky coffee tree," reads one report. "Seeds poisonous to humans, but seldom fatal."[21] Gymnocladus was awarded its common name by early white settlers in Kentucky, who used the roasted beans as a coffee substitute. As with true coffee (caffeine is an alkaloid toxin), the poison is in the dose. Too much of the bean will prompt the body to attempt a cleansing through vomiting and diarrhea. Irregular pulse leading to coma will follow, accompanied by a "narcotic-like effect."[22] (Perhaps there was more to this Kentucky brew than just the taste.) Saponin toxins within the seeds of the Chinese species of *Gymnocladus* have been found to inhibit replication of the HIV virus.[23]

"Tastes just like coffee," Carl Mehling told me. Carl is a paleontologist and educator at the American Museum of Natural History, and he enjoys collecting and eating edible wild plants. "I've served it to friends, and nobody ever suspected it was anything other than coffee. Even when I tell them, people have a hard time accepting that it's not the real thing."

"I've tasted the pulp," I responded, "but never the seeds."

"What? I've read the pulp is poisonous."

"I spit it out, of course. But it tastes just like the pulp of honey locust to me."

"I've tasted honey locust too," Carl added. "It tastes good."

"I've never been brave enough to try the seeds of gymnocladus though," I continued. "Seeds are usually even more poisonous than pulp. You actually use them for coffee?"

"Well, yeah."

"Does it feel like there is caffeine in the brew?"

"No."

"What about a narcotic?"

"No," he laughed. "It just tastes good. I roast the seeds first, for three hours."

"Oh, I'm sure roasting destroys a lot of the toxins—just like roasting does for coffee." I later learned that the toxins in gymnocladus seeds include impostor amino acids that substitute for the standard varieties animals use to build proteins.[24] A protein molecule that incorporates an impostor will function poorly if at all. High heat will destroy the amino acids, thus making the seeds safe to brew.

Carl Mehling had initiated this conversation when he tried to squeeze past my camera equipment that was perched over a mastodon jaw in a corridor of the vast room in which proboscidean fossils are stored (away from public view) at the museum. I had positioned a maclura fruit on one molar and dropped several gymnocladus beans into the valleys between massive horizontal ridges on the tooth's surface, trying to envision what elephantine grinding might have accomplished. This behavior would have looked strange to the uninitiated, so after finishing our coffee discussion, I told Carl about ghosts.

"What a great concept!" he exclaimed.

Like the honey locust genus, *Gleditsia,* the genus *Gymnocladus* seems to have originated during the Oligocene, about thirty million years ago. Around this time and for millions of years after, a land bridge connected Asia with North America. World climate was warm enough during the Miocene for both genera to inhabit the northern region of the Northern Hemisphere. Both honey locust and gymnocladus are careful to avoid exposing young leaves to frost. Gymnocladus is the last North American tree to leaf out in the spring, and it secures its buds from the biting winds of winter by retaining them within the twigs. Both qualities contributed to the scientific naming of this genus: *gymno* is Latin for "naked" and *cladus* means "branch."

Mature females are not, however, naked of all adornment during the winter. Gymnocladus pods hang on even longer than do honey locust pods. Many of the fruits remain on the branches well into spring. Because gymnocladus is a medium to large tree, even the biggest proboscideans would have had to wait many months for the pods to fall. The food value is not diminished during this time. The pulp does shrivel when it dries, but as I discovered in the fruit flotation experiment, the mass readily reconstitutes in water. The pulp I tasted was from a year-old pod that had been soaking in a bucket in my kitchen for two weeks. I changed the water daily during that experiment, because the saponins oozing out of the gap between the valves of the pod gave the water a silky, soapy feel.

What is the pulp for, if not to entice a big herbivore to swallow the seeds? Pulp is not de rigueur in a legume pod. North America's black locust (genus *Robinia*) produces flat, pulpless pods with papery hulls that open along both edges, releasing seeds while the pods still hang on the tree. The redbud (genus *Cercis*) produces thin-hulled pods that are plucked from the branches and dispersed by winter winds. One might suppose that pulp offers gymnocladus a chance to retain spring moisture long enough for the thick-coated seeds to germinate. Unlike honey locust, the pods of gymnocladus develop a gape along one edge, allowing water to enter, so such an adaptive explanation is plausible. But there is a problem: the seed coat is so resistant that even after three months of soaking in my kitchen, the seeds had not begun to swell. Yet when I cut through the coat with a serrated knife, the seeds swelled and then germinated in just three days. A mastodon may have achieved the same effect by grinding the pod between molars and then subjecting the contents to gastric acids. As we have seen, once a plant lineage goes the route of enticing herbivorous mammals to carry its seed, it may come to depend on them to scar the armor acquired for the rough journey.

In mid-April, several months after our encounter in the proboscidean room of the museum, Carl Mehling led me to the spot in Central Park where he planned to collect freshly fallen pods. He had run out of faux coffee from the previous year's harvest. Not only were gymnocladus pods abundant beneath the tree, but at least as many were still aloft.

"There goes your dispersal agent!" Carl pointed at a frisky dog who had just picked up a pod and run off with it, a second dog taking up the chase.

"I don't think a wild canid would do anything so ridiculous," I huffed.

Examining the ground beneath the canopy, we discovered that trampling and mowing had liberated a good many seeds from the previous year's crop. Most such seeds had become shiny marbles, but a few that had been squashed into the soil had taken in water and swelled to a dull brown. None were rooting, however, and we saw no seedlings or saplings in the vicinity. Venturing into the brush ten paces from the trunk of the tree, we found a number of rotting pods amid the leaf litter. That's where I did my collecting, for I was thrilled to experience what Dan Janzen had called "the riddle of the rotting fruit."

There was no telling how old these pods might be—at least a year, possibly two or three. Almost all that I opened were even more rotten inside than out. Mold coated the seeds as well as the interior of the valves. Some seeds were so far gone that in their stead were powdery hummocks of greens, grays, and browns. Here in this squirrel- and rat-rich park was confirmation that the seeds of gymnocladus were unappealing to rodents. Here, too, was confirmation that, unless liberated from its pod, a gymnocladus seed is doomed to decay.

Before leaving the tree, Carl Mehling and I had one more task. He pulled out of his pack a bag containing two large fragments of mastodon molars. For a few minutes, gymnocladus and its long-gone mutualist were reunited.

AS WITH OSAGE ORANGE, it is helpful to look for an Old World analog for Kentucky coffee tree. An African legume pod, *Tetrapleura tetraptera*, is a good match. *Tetrapleura* is relished by megafauna, including elephants, but it is poisonous to people. In Africa, woody pods such as *Tetrapleura* contain spherical seeds that readily roll between elephant molars, promoting scarring rather than fragmentation.[25] Gymnocladus has not only the largest but the most spherical seeds of any of North America's candidate anachronisms. A 1935 description of the seeds is illuminating: "The very hard shell of the seeds is easily polished by friction; boys like to carry them as pocket pieces, for, constantly rubbed by the other contents of their pockets, they become as smooth and shiny as glass marbles."[26]

Rare, usually found in floodplains though quite capable of growing in uplands, ignored by wildlife—not consumed even by cattle or horses—and way overbuilt for gravity dispersal: these characteristics suggest the loss of animal mutualists. Gymnocladus fruits are not simply anachronisms; they are anachronisms in the extreme. Small animals cannot begin to make up for the loss of the big guys because nobody, absolutely nobody, disperses the bulky fruit of gymnocladus.

Three compensatory life history traits may have been crucial for the survival of this tree during the past thirteen thousand years: poisonous seeds, poisonous leaves, and root-suckering skills. The seeds of gymnocladus are so poisonous that, in contrast to the seeds of honey locust, they are not attacked by bruchid beetles.[27] Whereas the seeds of more vulnerable legumes must vacate the vicinity of the parent in order to escape seed-predatory insects, gymnocladus seeds can remain where they fall. The downside of this extraordinary repulsiveness is that, as I observed in Central Park, the hoarding services of rodents are compromised too.

On balance, it is hard to say whether the repulsiveness of gymnocladus seeds has worked for or against the persistence of this lineage in North America. The seeds are free of pests, but they attract no backup dispersers. There should, however, be little doubt that the extreme poisons in the foliage of gymnocladus have helped this genus persist. The leaves are virtually pest-free. Mammalian browsers are likewise repelled by the toxins, and farmers are advised to keep their livestock away.

Finally, a third compensatory life history trait is one that causes trouble for arboretum managers, including Peter Del Tredici. "If you had come to the arboretum last year," he told me, "you would have been amazed by the number of root suckers surrounding our *Gymnocladus* specimens. But I've been on the case of the crew to clean them out."

Peter Del Tredici is director of Living Collections at the Arnold Arboretum of Harvard University in Massachusetts. That means he is in charge of the botanical specimens on the grounds and in the greenhouses, as distinct from the herbarium collection and the library. I had come to Arnold Arboretum in late May of 1999 to interview him about anachronisms. He offered to take me on a tour.

Dressed in casual pants and running shoes, Del Tredici was distinguished from the grounds staff we stopped to chat with only by his graying hair. A half dozen of his crew were riding in the back of an arboretum truck that looked a lot less beat up than the one he was driving. Our first stop was the gymnocladus specimens. I had noticed this grove of three mature gymnocladus trees on the walk into the arboretum from the subway station. Because the trees are surrounded by mowed lawn, it hadn't occurred to me that there would be any root suckers to examine. The closest patch of unmowed herbs and shrubs is almost ten meters away.

"I think I can see some suckers," Del Tredici stopped the truck to point out, "coming out of that bed of Xanthorrhiza." We walked toward the spot. "Usually they're over there," he motioned to the left, "but I see that the bed's been cleaned up. They'd be all over the place if we didn't mow." A few paces later he stopped. "There you go. See those two stems? Those are root suckers."

"Oh my god—from whom?"

"They're either from that one over there or from this one. I can't tell you which."

"Wow! That's great! Will *Gleditsia* do that too?"

"No. *Gleditsia* doesn't do that."

"Will *Maclura* do that?"

"*Maclura* does that, and so does *Robinia*. When you pull up a sucker, you find they have no root system under them. They're just attached to a long, lateral root. The suckers don't have a fibrous root system associated with them at all."

The tour continued. We visited *Maclura, Gleditsia,* and pawpaw.

"Here's the pawpaw. This whole area was infested with pawpaw—all from root suckers. I had to clean this all out. We took out over two hundred stems here. The original stem had died. That released the root suckers to just take off. We left these two. We'll see what happens."

His experience with the arboretum pawpaw substantiates what I've only been able to read about. But on one important point, Del Tredici presented me with counterevidence. Recall from Chapter 4 that pawpaw ova are rarely fertilized by pollen borne on flowers from the same clone. "Oh yeah, we get lots of pawpaw fruit here," he said, "even with just this one clone—it's remarkable."

The Tree Who Remembers the Dinosaurs

We didn't have time to tour the arboretum's ginkgo collection. I've seen lots of ginkgos in New York City, so it was more important for me to be shown less familiar trees. The ginkgos at Arnold Arboretum that would have interested me most are among the youngest in the collection, as these are some of the only wild trees left in the world.

A wild tree, native only to China, growing in a U.S. arboretum? The idea seems absurd because it is difficult for us to regard individuality in any way other than our own. An individual cannot possibly be alive in two places at the same time, no? From a plant's perspective, it well can. A fresh root sucker of gymnocladus is a distinct spatial expression of the same individual whose elder stem a few meters away produced it. And the pawpaw tree that died at Arnold Arboretum is not really dead. It is alive and well in two new stems that began as root suckers. The great bulk of this individual resides in the rootstock. If the growth of new leaves each spring is easy to regard as seasonal renewal of the same tree, it should be a small step to treat the cycle of birth and decay of clonal stems as renewal too. Yet our surface chauvinism blocks this insight. We forget the subterranean world ruled by roots because it is a world we do not inhabit.

An old ginkgo tree in China. This multi-stemmed tree is one of a few hundred individuals still growing in the wild. The Tian Mu Shan reserve has been set aside for their benefit. The rootstock from which these relatively young stems are growing may be thousands of years old. Courtesy of Peter Del Tredici.

It is, likewise, one short step further to extend the notion of individuality to encompass tissue transplanted to the other side of the world. Peter Del Tredici has done that for ginkgo. In 1989 he conducted field research at the Tian Mu Shan ginkgo reserve in a remote, mountainous region of eastern China.[28] This small reserve of natural forest contains the only (possibly) wild ginkgos remaining in the world. Even here ginkgo is a troubled plant; Del Tredici was unable to find a single seedling. Some young stems did exist, but they had been produced by clonal growth of old trees. Part of ginkgo's problem may be a paucity of suitably open sites for seedling establishment, but Del Tredici believes that other factors are important as well.[29] He returned to the United States with vegetative material cut from the old trees at Tian Mu Shan, from which he then coaxed new stems and roots for planting in Arnold Arboretum. In a very real sense, wild Chinese ginkgo trees are now growing partly in an alien land, and anyone who lives in Boston can easily visit them.

The land is not so alien, actually. *Ginkgo biloba* is an exotic in North America only if one insists on a constricted view of time. Rather, the tree has been helped by humans to reinhabit its native range. Fossil evidence shows that the genus *Ginkgo* thrived in temperate regions throughout the Northern Hemisphere during the Jurassic and into the Cretaceous. The species of *Ginkgo* alive today may have originated a hundred million years ago. It is truly what Darwin called a living fossil.

Ginkgo disappeared from North America just seven million years ago, and from Europe two million years ago. The most recent fossils come from southwestern Japan, not China. Mountainous regions are poor environments for preserving organic material or leaf impressions. Erosion rather than sedimentation is the geological condition that prevails in steep terrain. Thus the ginkgo lineage's relict population held on in a region that would have produced no recent opportunities for fossilization.

Whether or not the ginkgos at the Tian Mu Shan reserve truly are wild is a topic of debate. The tree has been associated with humans for so long that the question resists an answer. Perhaps the trees dwindled since the Mesozoic to just this one spot. Or perhaps the last wild population existed elsewhere, but the forest was converted into farmland. The trees living in the mixed-species forest at Tian Mu Shan today may be the offspring of trees originally planted by monks in the vicinity of an old temple.[30] Raw ginkgo seeds have long been used medicinally and as food when cooked (cooking inactivates most, but not all, toxins). The tree is also valued for two reasons beyond the material. First, because the roots, leaves, and wood are resistant to insect and fungal attack, ginkgo stems can live for several thousand years, attaining an enormous girth.[31] Very old trees are

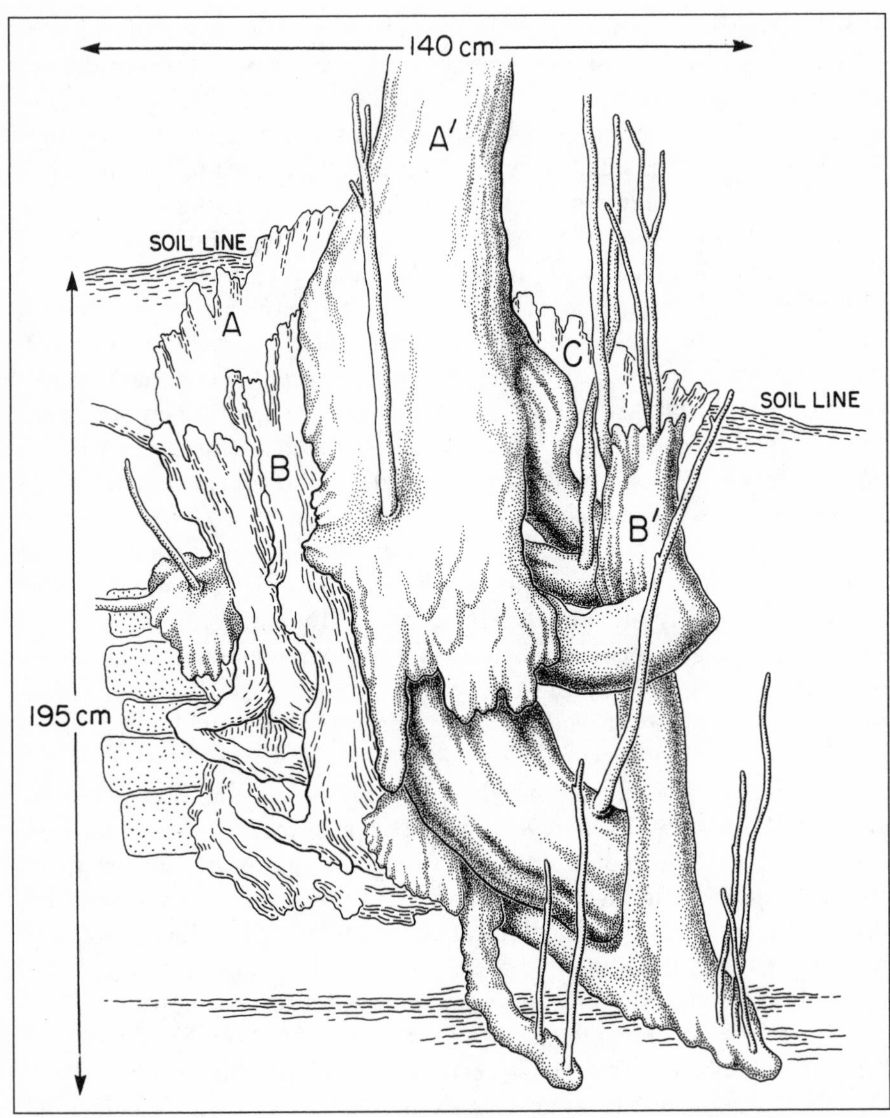

AN UNUSUAL FORM OF CLONAL REPRODUCTION IN GINKGO. Basal chichi from the remains of cut stems (A, B, C) have developed into new stems (A', B'). Courtesy of Peter Del Tredici.

revered in Asian cultures, and the oldest specimens in any given area are always found growing in association with Buddhist or Taoist temples. Second, ginkgo is an unusual tree in that it prepares for clonal rebirth above ground as well as below. Strange stalactites appear on the trunk and older branches, especially of ancient trees long past their reproductive prime. In Japan these growths are called *chichi,* which means "breasts." Tradition holds that nursing women can improve their flow of milk by making a pilgrimage to a ginkgo tree that bears prominent chichi.[32]

It was aerial chichi, not root cuttings, that Peter Del Tredici harvested from Tian Mu Shan reserve. Ginkgo does not produce root suckers; it cannot grow new stems from any portion of its root system. For clonal persistence, ginkgo relies instead on aerial and basal chichi, which are both essentially downward-growing stems. Composed of suppressed buds, chichi build up stores of nutrients and energy from which to send forth both roots and shoots, should the opportunity arise. Aerial chichi will kick into action when they touch ground—as when an old trunk falls or a human partner removes a chichi for planting. Not every ginkgo is old enough to produce such aerial growths, but all produce basal chichi. These emerge from the lowermost cells of the stem of every plant newly germinated from seed. Basal chichi will sprout a fresh stem immediately if the main stem is injured during early development. Basal chichi not called into early service for vertical growth nevertheless keep growing. But most of that growth of suppressed buds takes place underground. When this happens, the structure is called a lignotuber. Storing enormous quantities of energy and nutrients, a lignotuber is not simply an adaptation in waiting, one that will ensure longevity of the individual genome. Oozing around rocks it becomes a clasping organ, offering a solid hold for the existing stem.

Ginkgo is not alone in producing lignotubers. Sequoia, olive, and eucalyptus produce them too. In sequoia country along the coast of California, lignotubers are usually called burls. Del Tredici showed me an old photograph of one such burl exposed by beach erosion. The solid mass looked about as big as my New York apartment.

"You can see maybe eight trunks, some with three-foot diameters, attached to this burl. The original one in the center must have rotted away a long time ago," he explained. "I've done a lot of research on *Sequoia sempervirens,* subsequent to my work on ginkgo. Both species have this swollen base, which not only has the ability to produce shoots when the tree is injured or under stress, but it produces roots very readily. And that means that these suckers are capable of escaping the limitations imposed by the old root system, the root system of the parent tree. They generate a whole new root system, as well as a shoot system—whereas for an oak, you can get a new shoot out of the base, but they never generate new roots. So they're always tied to the old root system. Eventually the diseases catch up with the roots of the oak tree, and the whole thing—the clone—dies."

Peter Del Tredici knows more about the clonal and longevity powers of ginkgo than perhaps anyone else. His experience derives from field obser-

THE LARGEST GINKGO FOREST SINCE THE PALEOCENE. The trees in this ginkgo plantation in South Carolina are cut back to ground level every fourth or fifth year to allow easy harvesting of leaves from sucker shoots. The leaves provide a medicinal extract for the herbal health market. Courtesy of Peter Del Tredici.

vations and greenhouse experiments with aerial chichi, basal chichi, and basal chichi that have developed into lignotubers.[33] The understanding he has gained has had practical benefits for managing and expanding the collection of living ginkgos at Arnold Arboretum, including the dwarfed ginkgo specimens in the bonsai collection. Some of the work was sponsored by a French pharmaceutical company that finds profit in maximizing leafy growth at its ginkgo farm in South Carolina, which, Del Tredici believes, "may be the largest ginkgo forest since the Paleocene."[34] That forest is a monoculture of cultivated ginkgo plants that are cut back to ground level every fourth or fifth year. More than ten thousand trees grow in rows three feet apart on a thousand-acre site.

The herbal capsules available in the United States probably contain leaves from those trees. The Chinese did not use ginkgo leaves for medicine, perhaps because its active ingredients (terpenoids and flavonoids) must be extracted and concentrated in order to be effective. The uncooked seed has long been used to treat lung or kidney ailments, but nothing was made of the leaves. Nevertheless, ginkgo leaf's salutary effect on the human brain has recently been confirmed by scientific studies.[35] The results are strong enough that I decided to give ginkgo a try—out of fear of Alzheimer's disease, which has plagued one side of my family.

Peter Del Tredici got interested in ginkgo as a research focus not because of the plant's extraordinary cloning capacities and not because grants from pharmaceutical companies could be obtained for such research. He had no idea his ginkgo work would lead in that direction. Rather, he began to study ginkgo after encountering the anachronism paper by Dan Janzen and Paul Martin.

"I read that paper when it came out," he told me during the interview in his office. "It had a big influence on me. I had just been to Costa Rica, so I knew what they were talking about—the idea that the way things are today has very little to do with how they once were. All of a sudden, all these things I had been seeing made sense—like honey locust suddenly made sense for you. Before then, I didn't have the intellectual framework to make sense of what I was seeing."

He continued, "I was also very pleased with that paper because they missed it on ginkgo. They mentioned ginkgo, but they didn't quite get its story right."

What did Janzen and Martin say about ginkgo in 1982? In the final paragraph they suggest that the anachronism idea can be of service for understanding fruits beyond the neotropical limits of their own study. They speculate, "The vesicatory ripe fruits and weak-walled nuts of *Ginkgo biloba* might even have been evolved in association with a tough-mouthed herbivorous dinosaur that did not chew its food well."

Library research on ginkgo, along with discussions with vertebrate paleontologists at Harvard, led Del Tredici to a different conclusion. Yes, the "fruit" of ginkgo (*diaspore* is the proper terminology for all conifer propagules) is an anachronism, he concluded. But the plant wasn't trying to woo a dinosaur to swallow the fleshy fruit and defecate a viable seed. Rather, ginkgo was attempting to lure some of the earliest mammals—the multituberculates—to extract the seed from the oily flesh and bury it for later consumption, just as squirrels hoard nuts in temperate forests today. Multituberculates are sometimes called "the rodents of the Mesozoic." The multituberculate Del Tredici thought most squirrel-like in size and bone structure (including adaptations for tree climbing) was *Ptilodus*.

In his view, the flesh of a ginkgo fruit was meant to repel, not attract—much as the fibrous, oily husk of black walnut repels fruit-eating mammals today. It is then up to deft little creatures to extract the prize within, caching or burying much of the harvest for consumption during lean months ahead. Some of the hoarded seeds will never be retrieved, and some of the unretrieved may be situated favorably enough to germinate and establish. Del Tredici concluded in a 1989 publication, "In their adaptation to multituberculate dispersal, *Ginkgo, Torreya,* and *Cephalotaxus* [all gymnosperm trees bearing big-seeded, fleshy diaspores rather than dry cones] are evolutionary anachronisms, analogous to the Central American angiosperms that lost their original mammalian dispersers about ten thousand years ago."

His views changed following field experience at the ginkgo reserve in China. He still thinks that multituberculates were potentially important dispersers of ginkgo seeds, but he now suspects that dinosaurs probably played a role as well. The ginkgo associate was not, however, the kind of dinosaur that Janzen and Martin proposed. "It isn't what you would think—herbivorous dinosaurs. It's carrion-feeding dinosaurs—scavengers."[36]

While in China, he learned that red-bellied squirrels *(Callosciurus flavinmanus)* extract ginkgo seeds from the oily flesh for hoarding—just as gray squirrels *(Sciurus carolinensis)* are known to do in North American parks where ginkgo has been planted. Residents of the area also told him that small carnivores sometimes consumed the whole fruit, flesh and all. Among the alleged dispersers were the masked palm civet (*Paguma larvata,* family Viveridae), the leopard cat (*Felis bengalensis,* family Felidae), and the raccoon dog (*Nyctereutes procyonoides,* family Canidae). The palm civet is highly frugivorous, has a wide gape, tends to swallow fruits whole, travels long distances, and defecates seeds intact. The name refers to its habit of eating palm fruits. It is also partial to figs and the large, fleshy fruits of family Annonaceae (which, in the Western Hemisphere, includes American pawpaw). The raccoon dog, a close relative of North America's gray fox and about the same size, is highly frugivorous too. Unfortunately for ginkgo, all these Asian carnivores tend to use their feces to mark territory rather than scattering dung throughout.[37]

Del Tredici did not have an opportunity to confirm by direct observation the taking of ginkgo fruit by carnivores, nor to examine feces. But the reports made him realize that the multituberculate "rodents" of the Mesozoic, like the rodents of today, were probably second-string dispersers of ginkgo. Their habits may not have exerted the primary selective forces that shaped the ginkgo fruit. Unlike the protective husk of a walnut, the repulsive flesh that surrounds a ginkgo seed may not be repulsive to all. The first set of target dispersers could have been small scavenging dinosaurs that lived during the Jurassic or Cretaceous. As Del Tredici sees it, the odor of a ginkgo seed, after it has lain on the ground for a few days, may mimic rotting flesh well enough to attract scavengers of all stripes.

Pegging partnership on carrion-feeding rather than plant-eating dinosaurs solves one very big problem. The seed is physically protected by a shell too thin to withstand contact with grinding machinery. A ginkgo seed may look like a pistachio nut, but the shell that protects it is fragile.

Herbivorous dinosaurs of the Mesozoic had two kinds of machinery that might have crushed a ginkgo seed. Teeth would have been an obstacle in some, and for the rest, a stone-filled gizzard would have threatened even more injury.

Those of us who do not live on farms are likely unaware of our ignorance about gizzards. Because we have encountered gizzards in the packages we extract from plastic-wrapped chickens, we think we know what the organ looks like. Those gizzards, however, not only have been sliced open, losing their spherical shape, but are missing a crucial part of anatomy, the lining. A few years ago I helped a neighbor kill and clean chickens. Upon opening the first gizzard, I was astonished to find a lining that resembled coarse sandpaper—and that could easily be peeled away from the thick muscular walls of the organ. In contrast to seed-eating chickens, highly frugivorous birds such as Australia's cassowary possess soft-lined gizzards out of necessity: some fruits the cassowary consumes have seeds so toxic that the bird must ensure those seeds do not rupture during transit.[38] How might a cassowary react to a reeking ginkgo fruit? No one has performed the test.

Recall that Janzen and Martin proposed ginkgo as an anachronism suited for a "tough-mouthed herbivorous dinosaur." It is widely known that the flesh of ginkgo fruits can produce a human skin reaction on contact, much as poison ivy does. If so, how can palm civets, leopard cats, and raccoon dogs consume ginkgo fruit? Three plausible explanations come to mind. Perhaps the mouth can handle what bare skin cannot (recall my experience with the latex of maclura fruit). Or perhaps the irritants are concentrated within the flesh rather than upon it, so an animal is safe if it swallows the fruit whole. Finally, maybe the fruit gains or loses toxicity the longer it ripens on the ground.

I know from experience that ginkgo fruits do not smell foul when freshly fallen. They don't have much of an odor at all. It takes a few days for butanoic and hexanoic acids to develop in the ripening flesh. Similarly, might the compounds responsible for skin irritation also intensify with time? To begin to answer these questions, I performed a crude test on the tame squirrels in a park near my apartment in New York City. Washington Square Park hosts twenty-five mature ginkgo trees, half female, and a huge population of tame gray squirrels dependent on bread and nuts provided by their human friends. When not begging for food, the squirrels are busy being squirrels. In autumn that means collecting and burying acorns and ginkgo seeds. The squirrels do not have all the ginkgo seeds to themselves, however. They must compete with Chinese Americans who come to the park equipped with long poles for knocking down fruits still hanging in the trees. Every such harvester I have encountered has been wearing surgical gloves.

For freshly fallen ginkgo fruits, the fleshy jackets the squirrels leave behind show one neat incision. Within a week, those remains begin to reek. After the 1998 harvest was over, I found a few ginkgo fruits along the curb where squirrels do not forage. Some of the seeds had been squashed out of their skins by foot traffic and were lying loose. A few evil-smelling fruits were still plump and whole. I collected one whole, stinky fruit, plus two seeds that were dry and free of flesh and one that I squished out of a stinky fruit myself. I then went looking for squirrels. I began by tossing to a squirrel one of the dry seeds. The animal took up the seed with no hesitation, moved a few feet, then quickly buried it. A second approached while the first was digging, so I offered the new recruit the second dry seed. Squirrel number two shelled and ate the seed as I watched. Both squirrels were ready for more action when I offered the moist seed with its varnish of flesh. The more assertive squirrel took the seed in its mouth and ran off with the prize. It halted a second later, however, dropping the seed. Immediately, the squirrel began to rub his muzzle vigorously with his front paws for what seemed an unusually long time. Both squirrels then ignored that seed.

I walked to another part of the park, where I offered a whole, vile ginkgo fruit to six different squirrels in turn. All were tame enough to attempt to crawl up my leg. Three did no more than sniff the fruit. The remainder rolled it around in their front paws before dropping it and resuming their begging. I began to see a pattern.

The same week I tested the squirrels I also initiated experiments on myself. I am very susceptible to poison ivy. Would I develop a rash from ginkgo? I began by drawing circles on my left lower arm, and rubbed the outer skin of a fruit on one patch and the moist inner pulp on the other. I waited several hours before washing both areas with soap. No rash developed. The fruit I used for that first experiment was freshly fallen. It had a faint odor that was strange and not at all inviting but far from nauseating. Three days later I repeated the experiment, this time using the skin and flesh of the smelliest fruit I could find in the park, drawing circles on different patches of skin on the same arm. Somewhat more than twenty-four hours later, the circle that had been rubbed with moist pulp began to redden, while the patch rubbed with the exterior showed no reaction. The rash became itchy, but it did not advance to blisters. After three more days it began to subside. I concluded that while this particular human does experience contact dermatitis with the inner flesh of ginkgo, I could not judge whether the rancid fruit or my own heightened sensitivity from repeated exposure was responsible for the rash on the second try.

I should have ended the experiment there, but I decided to do one more rubbing. Again, I developed a skin rash only from the moist flesh. But before the rash subsided, other things started to go wrong. Two days after the rubbing I awoke with swollen eyelids. The next day my condition worsened, with my whole face puffy and tongue beginning to feel weird. I had never before experienced this sort of whole-body reaction to an allergen, so I called my sister, who has an asthmatic son. "If your throat begins to close up, you're in trouble," she warned. "Better go buy some Benadryl right now." Even with the Benadryl, three weeks passed before I began to look and feel normal.

I believe it is significant that only exposure to the inner flesh of ginkgo triggered my allergic reactions. Recall that the squirrels harmlessly rolled in their paws pungently ripe ginkgo fruit, while the squirrel I tricked into touching remnants of moist flesh appeared to suffer from the experience. My hypothesis is this: ginkgo flesh may pose little threat of contact dermatitis to mammals when the fruit is freshly fallen. But the interior becomes dangerous after ripening on the ground. It would be very interesting to know whether the palm civet, leopard cat, and raccoon dog in China prefer freshly fallen fruits.[39] Or, if they consume rank ginkgo, might they swallow the fruits whole?

Several months after my bout with ginkgo, my mate treated me to dinner at a restaurant specializing in mixed Asian fare. "I've got a surprise for you," he said. There on the menu were two selections containing ginkgo seeds. The entree I ordered was a delicate broth with various seaweeds and mushrooms, daikon radish, and about twenty ginkgo seeds. I knew that the toxins in ginkgo seeds are distinct from the acids in the flesh that had triggered my allergic reaction, so I figured it would be safe to eat the soup. Fortunately, I was right. The seeds were delightful—initially. Cooked ginkgo seed is sweet, meaty, but very rich. The richness became cloying, so I left a few uneaten. Even when cooked, the seeds are toxic, but the poison is in the dose. Some publications warn adults to eat no more than seven seeds at a sitting.[40] In China the people who die from eating too many seeds are usually small children. The toxins kill by interfering with the activity of vitamin B_6.

The next fruiting season I hoarded some ginkgo fruits in my freezer. These I intended to take to New Mexico the following summer—though not to try out on Mrs. Foxie. She would make an excellent, though gingko-naive, proxy for a Chinese raccoon dog, but there are limits to the scientific quest. In this case, bonds of affection trumped curiosity. No, these fruits were not for foxes but for dinosaurs—carrion-eating dinosaurs.

Twenty-some turkey vultures sleep in the cottonwood trees a short way upstream from us, but at dawn they fly to the ponderosa pines and cliffs on our property to bask in the sun and catch the first thermals that rise from this east-facing wall of the canyon. The vultures have grown accustomed to my visits, so they simply watched as I placed beneath one of their trees a shallow container with ginkgo fruits. I watched from a distance until all the birds had circled away for the day: not one had investigated the novelty. Three days later the fruits (now desiccated) were still there. Maybe vultures are suspicious of the unfamiliar sight or smell of ginkgo. Turkey vultures find carrion more by smell than by sight. Or maybe they wouldn't feel safe on the ground right there. So I placed another set of fruits, sans container, on a gravel bar where I've seen vultures feeding before. They circle that spot repeatedly every evening, so they should be able to detect the odor. Alas, there, too, the fruits were ignored.

Does this negative or null result refute the dinosaur hypothesis? Surely not—though, I admit, had the vultures gobbled up ginkgo I would be talking confirmation. So what can we say? Is the fruit of ginkgo anachronistic? Might the thin shell surrounding the seed and the chemical features of the oily flesh be adaptations of another era, out of step with our own? My dabblings with turkey vultures, squirrels, and allergic responses are suggestive, but far from conclusive.

Peter Del Tredici is the first to admit the weakness of the evidence on which the dinosaur hypothesis is based. Still, nobody has offered a better-supported explanation for the mystery of ginkgo.

I would like to believe that ginkgo has been on the decline since losing its preferred cadre of ground-dwelling, carrion-feeding dinosaurs. I would like to believe that ginkgo still dreams of dinosaurs. And I would like to believe that this ancient (and now unique) lineage held on ever since through extraordinary powers of persistence—by its virtual immunity to insect and fungal attack and by the opportunities for rejuvenation offered by lignotubers. I would thus like to believe that we humans rescued from extinction the genus *Ginkgo* and thus the last remnant of a once mighty taxonomic order (Ginkgoales)—and that we did it in the nick of time. That story is lovely to hold in mind as I walk the ginkgo-shaded sidewalks of New York City. But is it science?

"The truth of the matter—at least from my perspective—is that there was no one cause for ginkgo's decline," Del Tredici tells me. "My twenty-five years of working with plants tells me there is never a single cause for a given effect. Consider: in addition to whatever its current dispersal problems may be, ginkgo has a terribly archaic system for producing seed. Pollination occurs in April, but fertilization doesn't happen until September. So the tree has to make a full-size seed before it even knows whether it's going to have an embryo or not. A seed is a huge investment, so that's a very inefficient approach.

"Ginkgo also has a very primitive sequence of development. The embryo has to grow while the seed is on the ground, whereas most plants shed seeds with fully developed embryos—the vast majority of angiosperms are that way. (Magnolias are an exception; they have a similar type of dormancy.) You've got to remember that the angiosperms started offering stiff competition fifty or sixty million years ago."

"I know ginkgo needs a lot of sun," I offer.

"Yes, it's a pioneer species. I call it a persistent pioneer. It needs sun to get established, and then it can get shaded out. But it will still persist until the surrounding forest gets hit by some form of disturbance. It's a pioneer that just keeps coming back.

"Seedling establishment is the bottleneck for ginkgo," he continues. "It's the most difficult thing for a tree to do—not to disperse a seed, but to establish a seedling. That's true especially for conifers, and gymnosperms in general. That's where they are most vulnerable to competition. Dispersal is a prerequisite to establishment, but to my way of thinking, dispersal is easier than establishment for a gymnosperm to accomplish. Gymnosperms are not as competitive as angiosperms in this regard. They have a much more difficult time because their seedlings grow more slowly than those of angiosperms. In general, the seedlings of gymnosperms are very conservative; they're slow. They grow a finite amount and then just stop. There's a famous article, "The Tortoise Versus the Hare," which looks at seedling regeneration patterns in angiosperms versus gymnosperms. In so many ways, the angiosperm seedlings are ahead.

"I think ginkgo, evolutionarily, is just way out there in left field," Del Tredici goes on. "The only reason it survived is because of its amazing ability to keep resprouting. It has increased its life span to the point where, given enough time, it can reproduce from seed. Increasing the life span of the individual increases the chances of seed success."

Whatever suite of problems ginkgo may have faced during the tens of millions of years before we came upon the scene, a delicate constitution is surely not to blame. Ginkgo may be the hardiest, most invincible tree on the planet. That is why it handles New York City so well. As Del Tredici has written, the ginkgo has "an unparalleled tolerance for environmental stress."[41] A shrine in Japan is a poignant tribute to the resilience of this ancient tree. The shrine calls attention to a particular ginkgo tree that would seem too young for recognition anywhere else in Japan. The stem resprouted from rootstock only a half century ago. But it stands in Hiroshima.

"I have a concept I call ecological immortality," Del Tredici tells me. "As long as conditions remain stable, there's no reason that a ginkgo individual can't go on living forever."

ANACHRONISTIC THORNS. Bare branches of four trees native to North America are shown here with roses for scale. FROM LEFT: Mesquite, osage orange, hawthorn, and honey locust.

7

Armaments from Another Era

Animals may be willing or unwilling partners in seed dispersal. Fifteen thousand years ago, gomphotheres gorging on the fruits of avocado trees in tropical America would have been enthusiastic partners in a mutualistic relationship. The energy-rich pulp was the plant's part of the bargain; safe transport of seeds, the quid pro quo easily accommodated by an elephantine gut. The North American fruits thus far presented as anachronisms had all forged mutualistic arrangements with a succession of megafaunal partners. That tradition atrophied at the close of the Pleistocene epoch.

Replacement partners of lesser stature—notably, foxes and raccoons—made up for the loss in the case of persimmon. The same second-string partners also became important, though probably less effective, dispersal agents for pawpaw. Meanwhile, rodents with a penchant for hoarding have stepped in for the big guys who once partnered honey locust and desert gourd, but they make an odd match. Honey locust and desert gourd have dressed for the opera, but their dates insist on the movies. The fruits are way overbuilt for the only mode of dispersal the animal realm now can offer: extraction and caching of seeds. For North America's extreme anachronisms, nobody seems to have been waiting in the wings when mobile helpers vanished. Unless carried away by floods or a tumble downhill, the pod of Kentucky coffee tree simply rots where it falls. The fruit of osage orange rots too, though after squirrels have peeled away the seeds while sitting in safety at the base of the parent tree.

The dispersal strategies of persimmon, pawpaw, honey locust, desert gourd, Kentucky coffee tree, and osage orange all suggest bygone partnerships. In contrast, this chapter begins with a look at anachronistic instances of one-sided affairs. Here the animal is drafted rather than lured into seed dispersal service.

Burs that entangle in fur are fruits that offer no reward to the animals who unwittingly disperse them. Such relationships are commensal (harmless to the animal) at best and parasitic if the burs irritate or injure the host. In their 1982 publication, Dan Janzen and Paul Martin presented a list of Costa Rican plants that produce anachronistic burs. These burs are occasionally distributed by lesser beasts native to Central America, and some are currently well served by cattle and horses, but their preferred partners—the large mammals who exerted the strongest evolutionary influences upon them—are now ghosts. "With the loss of the megafauna," wrote Janzen and Martin, "we suspect that many of these plants declined severely in density and some even suffered local extirpation, as the once open and well-trampled habitats were reforested and as seeds were no longer dispersed by large shaggy beasts such as gomphotheres, toxodons, and ground sloths."

What makes these burs anachronistic is not an outright absence of fur in post-Pleistocene landscapes. Rather, it is a paucity of fur of the proper texture or at the proper height. Dan Janzen tested the adhesiveness of Costa Rican burs on the largest animals native to the post-Pleistocene American tropics. The burs failed to adhere to either tapir or peccary. That result led me to ponder whether the tall cockleburs (genus *Xanthium*, family Asteraceae) that grow in the river bottom near my home in New Mexico are perhaps better suited for the horse tails that have recently reappeared (after a thirteen-thousand-year hiatus) than for well-groomed foxes and coyotes or the sleek haunches of deer.

Cockleburs are now found from the pampas of Argentina to the plains of Alberta. Each bur is about the size of an almond and contains two seeds. Abundant spines decurve in pointed hooks—which (true story) became the model for Velcro. Cockleburs "transform the tails of cattle to prickly clubs and hang on, if not removed by man, until the hairs are shed."[1] I tested *Xanthium* on the fur of a roadkill deer I came upon while bicycling one evening, as hip-high stalks of *Xanthium*, though dead and dry for half a year, were still erect on the nearby floodplain. The burs all too readily entangle in my shoelaces and socks, and they even stick to my fingertips, but their grasp on the fur of the dead doe proved feeble.

How Devil's Claws Got Their Claws—and Other Stories of American Deserts

Two years after their joint paper on Costa Rican anachronisms appeared in *Science*, Dan Janzen sought Paul Martin's advice on anachronistic features of Chihuahuan desert plants. "If you don't know about devil's claw, you have missed the best of the lot," Martin wrote back.[2] The deserts of Mexico are not only close to Martin's home in Tucson but close to his heart. For

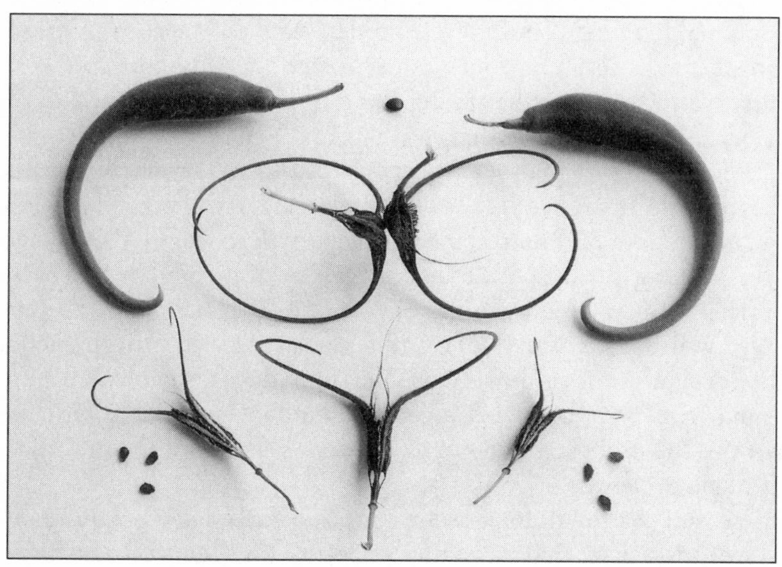

DEVIL'S CLAW. TOP: Two unripe fruits of *Proboscidea* with an assortment of opened claws and spilled seeds. (Honey locust seed, top center, for scale.) BOTTOM: Devil's claw in action. The human foot is not ideally structured to pick up devil's claw, but horses and cattle all too easily acquire these giant burs just above the hoof. Some seeds spill out readily; the remainder require trampling for liberation.

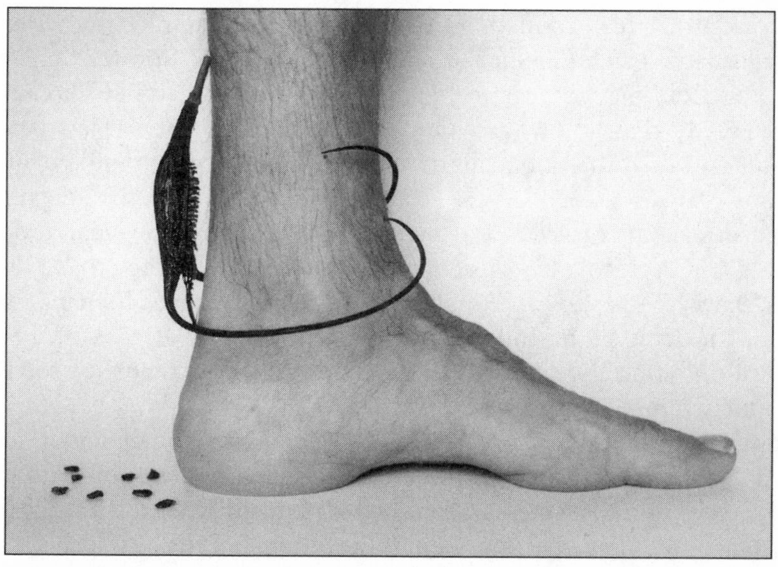

fifty years, he has been taking students and colleagues on scientific and instructional field trips south of the border. Now an emeritus professor, he no longer makes teaching forays into Mexico, but my book project gave him an excuse to revive the tradition. Paul, myself, my mate, and one of Paul's graduate students headed south in mid-September.

Before crossing the border, we stopped at a place where Paul could guarantee me a first encounter with devil's claw. This was a horse corral, and just as he had promised, *Proboscidea parviflora* (family Martyniaceae) was thick along the fence. Because of sticky and repugnant leaves and stems, this plant was virtually the only species that had not been eaten or trampled within the corral. Not yet ripe, the large green fruits looked more like their scientific than their common name; they resembled an elephant head and trunk. Paul assured me that after the outer layer browns and peels away, the proboscis portion of the woody black fruit would split into a pair of giant claws.

The season of growth for this annual plant must have got off to an earlier start in Mexico. Two days after encountering the unripe fruit at the horse corral, I was introduced to fully ripened devil's claw about a hundred miles south of the Arizona border. Our party had stopped to explore the vegetation along a dry wash. To my delight, devil's claws fully extended for action were hanging in abundance from knee-high plants. Walking carefully, I did not give any a chance to catch a ride. People who live in devil's claw country regard these plants as a nuisance for precisely this reason.

A year after my introduction to devil's claw, I began to watch and water a rare, isolated patch I happened upon a mile from my New Mexico home. One day trekking home after visiting the patch, I stopped to retie a shoelace. My right boot was encircled by an old bur—one claw beneath the sole and the other looped over the toes. After a quarter mile of abuse, the pod still contained a few seeds. Clearly, this adhesive organ was designed for appendages at least as thick as our own. The small foot and slim leg of a deer, coyote, or peccary slips right through the broad grasp of those claws. Domestic cattle and horses now join booted humans as fine proxies for dispersal partners who came before. Thus, after an absence of thirteen millennia, suitable carriers are once again available to serve the reproductive interests of *Proboscidea*.

Humans have unintentionally assisted the resurrection of devil's claw in North America. Thanks to livestock, this genus has expanded its range from Mexico and the American Southwest all the way up Iowa and over to Louisiana. For some centuries before Europeans arrived, indigenous peoples had developed a mutualistic relationship with the plant. The ripe fruits were gathered for the long, dark fibers that could be peeled off the claws and woven into baskets of pale yucca, beargrass, or willow to create distinctive patterns. Devil's claw was sometimes cultivated rather than just collected. Ethnobotanist Gary Paul Nabhan describes a cultivar of *Proboscidea parviflora* that is still used by traditional basketmakers.[3] Over the generations the claws have lengthened, and enhanced flexibility has been bred into the fibers. Somewhere along the way, the normally pink and purple flowers paled, as did the once dark seeds. The fibers of the domesticated fruits are still, however, strikingly black.

John Brewer, the same writer friend of mine who compared maclura to breadfruit in the previous chapter, reflects on his fascination with devil's claw: "It's such a savage-looking thing, with its surgically pointed claws, all dry and brittle. I was always a little scared of them as a kid—which gave them power."[4] Brewer urged me to seek out a sculpture of devil's claw the next time I passed through the Phoenix airport. Devil's claw is by far the biggest bur in North America and perhaps in the world; the sculpture at the airport is a celebration of size. The claws arc far over my head. Except for scale, every detail is precisely correct. Devil's claw is so strange that the maker, Kevin Berry, found no need to embellish upon nature in order to produce a dazzling work of art. The title: *Embrace*.

One other desert plant also attaches to passing animals in a big way, but it is not a bur; the adhesive organ of this plant contains no seeds. The mode of reproduction is vegetative rather than sexual. This plant is the jumping cholla cactus, *Opuntia fulgida*. Its spines serve a dual purpose. Like all cactus spines, they repel vertebrate herbivores; but outfitted with microscopic, backward-pointing teeth, they readily attach to a passing hide. Segments of the scrawny branches of this cactus detach at the slightest touch and cling to whatever did the touching.[5] So friable is the plant and so adhesive the spines that passersby swear that the branch itself jumped out to hitch a ride. Like many cacti, jumping cholla easily roots if a detached branch lands on favorable ground.

The seeds of jumping cholla are dispersed by animals too, but within their partners, not upon them. Cholla produces a sweet pulpy fruit containing many small seeds. The cactus seeds and skin that appeared in the coyote turd near my home (described in Chapter 5) were those of a jumping cholla.

One cannot help but think of megafauna when encountering the magnificent fruits of some of the larger subtropical species of genus *Opuntia*. Dan Janzen depicts these fruits as vivid reminders of a bygone era.[6] The

Cactus with spines and fruits adapted for missing megafauna. The fruits on this tall *Opuntia* are out of reach of native animals alive today. With texture and taste akin to watermelon and containing myriad small seeds, *Opuntia* fruit would have appealed to Pleistocene beasts as a source of water as well as energy. Honey locust seed (1 cm) for scale.

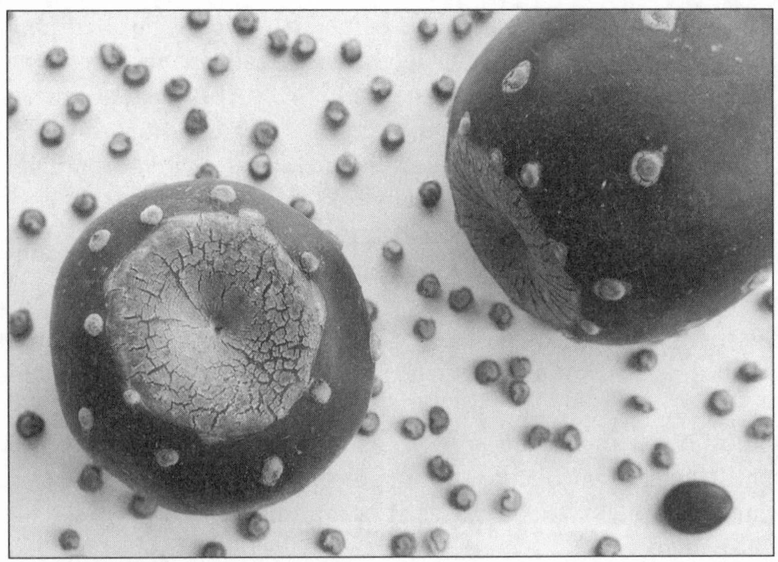

tallest species of the prickly pear group of *Opuntia* cacti bear fruits well out of reach of the tallest desert mammals alive today—though not of reach of the tallest desert mammals of thirteen thousand years ago. American camels and desert-adapted ground sloths would have had no problem harvesting fruits from even the tallest prickly pears. Gomphotheres probably made forays into the desert during the height of the fruiting season.

If prickly pear fruits promptly fell to the ground upon ripening, only their size would betray the ghostly presence of missing partners. Coyotes, foxes, peccaries, and tortoises are eager to consume fallen fruits and are quite capable of defecating intact seeds. Nevertheless, access is a problem because some cactus species retain ripe fruits for months. Eventually, a cactus abandons the wait. Its fruits desiccate and drop, no longer offering prospective dispersers a precious gift of water.

Prickly pear cacti may be anachronistic in yet another way. The tallest subtropical species bear spines at heights far above the browsing range of extant desert herbivores. Are those spines reminders of a former time?

Dan Janzen believes so. In 1986 he published a paper that carried the anachronism idea to the level of landscapes: "Chihuahuan Desert Nopaleras: Defaunated Big Mammal Vegetation." This was the piece that Janzen was working on when Paul Martin advised him to consider devil's claw. Although Janzen regards devil's claw as an anachronism, he devotes far more attention to the plant for whom the distinctive nopaleras landscape of the southern Chihuahuan desert was named. *Nopal* is a Mexican Indian term for prickly pear cacti with stems expanded into broad pads. Nopaleras are dense stands of prickly pear one to four meters tall. Other tall but less abundant species in nopaleras landscapes are yucca (family Liliaceae) and the legume trees acacia and mesquite. These plants also are physically defended against megafaunal browsers, and all produce fruit adapted for megafaunal dispersers. One treelike form of yucca produces a cluster of sweet and pulpy fruits that hang as high as three meters above the ground—waiting for a camelid or proboscidean to come along.

Janzen judges the nopaleras landscape to be "highly anachronistic." He writes, "Surrounded by thorny plants and edible fruits, one cannot avoid visions of herbivorous megafauna that recently occupied these habitats." Herbivorous megafauna would have shaped nopaleras landscapes in two ways—by dispersing cactus, yucca, and legume seeds in their feces and by preferentially nibbling and trampling competing plants defended by less formidable armaments. Various kinds of yucca are protected by stiff points tipping their long, narrow leaves. The armaments of some are so dangerous that city dwellers are urged either to landscape their yards with something else or to snip off the tip of each yucca leaf. The legume trees have defenses too. Acacia and mesquite bear on their branches widely spaced thorns, some as long as my little finger.

Janzen believes the armaments of prickly pear, yucca, acacia, and mesquite evolved in response to tall herbivores with big mouths. As demonstrated by giraffes, black rhinos, and elephants in Africa and by camels in Old World deserts, spines and thorns do not entirely shield plants from browsers, although they do reduce the damage. We cringe when a nature program on television shows a giraffe casually munching a thorny branch of *Acacia tortilis,* whose picturesque umbrella shape is, in fact, a joint creation of tree and mammal. Similarly, Janzen reports, "I have watched camels eagerly browse spiny prickly pear hedges in southern Morocco."

The cactus family is thought to have originated in South America, spreading later into North America. Prickly pear now thrives in all too many parts of Africa, Australia, and just about everywhere that rainfall is meager. We humans have carried it there. Our livestock then made sure that the introduced plant would feel at home. Old World camels now munch on New World cacti. The relationship, however, has deep roots. Camels originated in North America and spread into Asia, Africa, and South America long before humans began to intervene in the geography of life.

The mesquite genus, *Prosopsis,* likewise originated in the Americas. A minor component of nopaleras, mesquite is the dominant plant in some arid shrublands. Forty of the world's forty-four living species of mesquite are native to the Western Hemisphere, three-fourths of those to South America.[7] The pods and seeds are so nutritious and tasty that Pima and other Indians of the American deserts made wild mesquite a staple food. The sweet pods are ground into a meal, mixed with water, then dried into a cake that is eaten uncooked.

Mesquite is far more abundant today than it was before Europeans invaded the New World. Horses and cattle not only disperse the seed in their dung but also enhance germination by abrading the layers of armor surrounding the embryo. Also important is the service livestock provide in readying the landscape for mesquite seedlings, which have a difficult time establishing in untrampled grasslands.[8] Foxes and coyotes deposit viable mesquite seeds in their dung, but they do not abrade the seed coat or prepare the soil for germination. For all these reasons, mesquite fruits may be at least moderately (and perhaps substantially) anachronistic.

The recent robust spread of mesquite via horse and cattle dung is widely regarded as unnatural. From a Pleistocene perspective, however, it can be

Pods and seeds of mesquite. All mesquite species (genus *Prosopsis*) protect their seeds with both a tough seed coat and a bony case (endocarp). Samples of endocarps paired with seeds are shown for three species native to the Western Hemisphere. The specimen at top has had half its pod removed, revealing the layout of the endocarps—several of which display escape holes of bruchid beetles.

seen as a return to prehuman conditions.[9] Natural or no, the invasion of mesquite is now perceived as malevolent. For ranchers in West Texas and Argentina, the transformation of grasslands into mesquite shrublands threatens livelihoods.[10] Although mesquite pods are high-quality fodder, they are an ephemeral resource. Throughout much of the year mesquite is barren of foliage. Even when the trees are shrouded in green, the foliage is protected by toxins strong enough to deter browsing by cattle. For much of the year, therefore, a mesquite-dominated landscape offers domestic livestock little more than starvation. Fodder in a grassland, by contrast, is available year-round, and many grasses offer sufficient nutrition for body maintenance (if not growth) during the driest times.

Mesquite pods would have been excellent seasonal foods for Pleistocene megafauna. Camels and perhaps proboscideans might have utilized mesquite even during the podless months, because thorns would have slowed but not prevented browsing of branches by these thorn-adapted lineages. The cactus-dominated nopaleras would also become attractive seasonally. Janzen writes, "It is easy to visualize a herd of gomphotheres ranging into nopaleras to eat prickly pear fruits in the summer, moving into the oak forest to eat acorns in the fall, and then back into the nopaleras to eat prickly pear pads in winter." In a range now fragmented by property lines and fences, however, migration is not an option. Shrubs that replace grasses, no matter how plentiful and nutritious their seasonal offerings, must be judged pests.

Once established, mesquite is difficult to eradicate. It takes a tractor (or an elephant?) to topple trees, and the genus produces root runners at an early age that sprout profusely if a main stem is injured or killed.[11] Mesquite's substantial cloning skills would have served it well, of course, during the hundred-some centuries it had to wait for megafaunal mutualists to reappear.

We can look to Africa today to learn how megafauna might have affected the mix of grassland and woodland in Pleistocene America. Norman Owen-Smith reports that elephants have been reintroduced into the Hluhluwe Game Reserve in South Africa "with the aim of halting a steady worsening trend toward closed woodland and forest."[12] The elephants browse, debark, and push over trees invading grasslands. Presumably they would do the same to American savanna threatened by mesquite incursions.

The finest mesquite shrublands and the finest nopaleras landscapes today can be found where domestic livestock have intensively grazed—many say overgrazed—the range. Cattle and horses will consume almost any leaf and stem before tackling a spiny cactus, thus favoring nopaleras growth. The nopaleras landscape of Mexico would have declined severely at the end of the Pleistocene. Its vitality today can thus be viewed as a renascence of vegetative form.

Dan Janzen likes to think of ranch horses in the Americas as "a Spanish gift from the Pleistocene."[13] Paul Martin agrees. Reflecting on the decision by the National Park Service to remove feral burros from the Grand Canyon, Martin pointed out that Pleistocene remains of extinct equids (as well as of Harrington's mountain goat, Shasta ground sloths, and California condors) have been found in the canyon's dry caves. He argued that by removing the burros the Park Service had not eliminated an alien species but denied citizenship to a species attempting repatriation. "Having moved in [to the Grand Canyon] millions of years after the evolution and radiation of the Equidae, it is not clear . . . why mountain sheep must be regarded as more 'native' than the feral burros."[14]

More, Martin believed, burros would be excellent food for wolf and condor restoration in the canyon. In 1990 he wrote: "The presence of fossil condors and other extinct species in Grand Canyon caves suggests another, even wilder experiment for the future: restoration ecology. I call this view 'thinking like a canyon.' Canyons fear nothing, neither herbivores nor predators, neither native species nor aliens—only that alien human beings might sell a canyon short. What might the North American landscape and its inhabitants look like if the cast of large herbivores and carnivores included species equivalent not only to those seen by the first white explorers, but also those seen by the first Native Americans? Could wild burros provide a food supply for wolf restoration in or around Grand Canyon? When people begin to think like a canyon, a brighter day for many more kinds of large animals—wolves, condors, and burros included—must lie ahead."[15]

Paul Martin may sound like a radical environmentalist, but he cannot be so easily pegged. Shrub oxen were in North America, too, during the Pleistocene, and Martin tends to regard cattle as the return of the bovids. Yet for many of us greenies, cattle grazing on arid and semiarid public lands of the American West is one of the greatest of all environmental sins. During our botanical field trip to Mexico, I would look for the most godawful examples of land abuse by ranchers. Pointing with indignation, I would demand of Paul, "Okay, now is *that* overgrazing?" His answer was always the same. "Underbrowsed," he would calmly reply.

America lost its megafaunal herb- and shrub-eating browsers as well as its grass-eating grazers at the end of the Pleistocene. Only the grazers have returned—and they have been forced by humans to wrest a living not only from the grassy plains of their liking but also from arid shrublands where grasses are naturally sparse. To Martin's way of thinking, the ecological problems of the arid West today owe to the fact that dry-land ranchers are working with "the wrong species."[16] Years before he spoke up in behalf of feral burros, Martin advocated reintroduction of the right species to the American West. In 1969 he published an essay in *Natural History,* where he wrote: "Perhaps the long-lauded home where the buffalo roam is also the land where camel and eland should play. . . . In the perspective of the fossil record, one finds that Asian camels represent a lineage that has a far longer history on this continent than the American bison, a genus that arrived [from Asia] only in the middle of the Pleistocene many millions of years after the American origin of camels."[17]

What effect would browsing camels likely have on the arid West? Martin did not need to speculate. The experiment was run in 1857 by George Beale, an officer of the U.S. Army, who used camels as beasts of burden while laying out the Great Wagon Road through Indian Territory

AN "UNDERBROWSED" LANDSCAPE OF THE AMERICAN SOUTHWEST. A near monoculture of creosote bush *(Larrea tridentata)* offers horses and cattle poor grazing on this ranch in southern New Mexico. During the Pleistocene, American camels would have eagerly browsed this desert landscape.

from Texas to California.[18] To his delight, Beale discovered that camels could not only make do without water and hay during long desert treks but that they could carry such supplies for the cavalry horses. He wrote in his journal that the camels preferred to browse on mesquite and other thorny shrubs, ignoring even "the finest grass." He continued, "It is certainly very gratifying to find these animals eating, by their own preference, the coarse and bitter herbs hitherto of no value, which abound always in the most sterile and desolate parts." Beale was amazed to find that the camels could "keep fat" on a diet of prickly pear and creosote bush *(Larrea tridentata)*.[19]

American ranchers did not, of course, import cattle from Europe in order to restore arid landscapes to a Pleistocene norm; they brought in cattle as food. Camel meat may not sound appetizing to those of us who live in a land of beef, but Arab tastes are sufficient to support a small export market of camel meat from Australia. Might nonvegetarian environmentalists in the western United States provide a ready market for camel meat? Landscape restoration has now become a priority, and camels could work synergistically with cattle, just as browsers work synergistically with grazers in the savannas of Africa.[20] To turn back the invasion of American grasslands by mesquite, cactus, creosote bush, juniper, and pine, Martin maintains that the proper means of control is not herbicides and tractors but the reintroduction of browsing megafauna who would not be deterred by thorns and resins. We should therefore welcome back to its evolutionary birthplace the tough-mouthed, iron-gutted, Old World camel.

"Ha! I could see Bactrian camels up here!" laughs Bob Langsenkamp. I've known Bob for almost thirty years and have watched his career trajectory from field representative for the Wilderness Society, to civil service honcho at the New Mexico State Land Office, and now to ranch manager. The trajectory is not the fading green spiral it might appear. The 36,000-acre ranch Bob helps manage is owned by the Conservation Fund, which has pledged to work toward ecological restoration of this mesa an hour's drive south of Santa Fe. The Conservation Fund has no cattle of its own on the ranch. Rather, it grazes cattle owned by others who commit to using the time out for range improvement of public lands leased to them for grazing. One way to improve the range is to set fire to it. That means getting the cattle off first—preferably well in advance of the burn date so that tall, ungrazed grasses will carry a flame.

On my first visit to the ranch, in the fall of 1999, Bob gave me a working tour. That meant firsthand experience of one way to improve the range: killing trees. Killing trees is hard work. I soon discovered that I had the strength to tackle only ponderosa pines—and then, only those less than half my height. The pinyon pines bent too easily to the ax, and the thick basal growth on a shrubby juniper thwarted my efforts to connect with the stem. Bob and I were working that day in what once had been a vast, grassy meadow. Now tree invasion was turning meadow into forest.

The first trees I killed were offered a formal apology before the ax fell. Then the apologies were thought rather than spoken and soon not thought at all. Restoration requires feeling on the scale of a landscape. "Thinking like a mesa," Aldo Leopold and Paul Martin would have called it. Even so, this sort of action is futile over the long run, as Bob acknowledges. A few people with axes will not solve the "problem" of shrub invasion of the American West, which stems from a concatenation of causes, beginning with the loss of Pleistocene browsers, followed by overstocking of the range by cattle and overzealous fire suppression.

Might free-ranging herds of camels accomplish what Bob and I could not? Sitting with our sandwiches in the shade of an old juniper tree, we scrutinized the landscape from a camel's perspective. I concluded that, like me, the camels would be able to kill the youngest invaders, but the older trees would have to be taken by other means. The grasslands were too far gone. Bob cautioned that even with camels the mesa might be doomed. There are too many people living too close to the ranch, and those people are not willing to breathe the smoke of fires set for the health of the land. Controlled burns may be too restrained in scale and frequency for them to do much good.

An increase in winter precipitation coupled with the global rise in carbon dioxide may also be working against the grasses. Grasses thrive where the climate is too moist for desert but too dry for forest. An increase in atmospheric concentrations of carbon dioxide may help trees as much as a few extra cloudbursts; trees can keep their pores closed more of the time and still get the carbon they need to build sugars from photosynthesis. The less time trees spend with pores open, the less water vapor is lost to the air.

"So what's wrong with trees anyway?" I ask, unwilling to accept that a degraded environment is inescapable in this magnificent terrain. Except for our own selfish interests in wanting to grow a tasty form of meat, why should the land itself care whether the vegetation is tree or grass? Bob recites the answer I have heard before: in arid lands such as these, soil becomes vulnerable to erosion when grasses are replaced by trees.

It is all very confusing. Biodiversity protection is a no-brainer. It takes little moral discernment to grasp why we should allow to let live what was here before our own rapacious ancestors came on the scene, and it is only a tad more difficult for scientists to figure out whether something here now was also here then. But when we begin to toy with ideas about rearranging the landscape or returning to a long-gone era, there is no easy certitude.

How Hawthorns Got Their Thorns—and Other Stories of American Woodlands

A week ago I had my first really good look at a hawthorn. Surely I have walked past many hawthorns before, especially in woodlands of the eastern United States. We seldom notice, however, what we are not prepared to see, and hawthorn is far from showy. Barely more than a shrub, it cannot join the grand canopy of a forest. The leaves of many kinds of hawthorn are small and tend to brown rather than brighten in autumn.

The eastern United States is the center of distribution for the hawthorn genus, *Crataegus* (family Rosaceae). Species of *Crataegus* are native to disturbed forest and forest-edge environments in this region—just the sorts of places one would expect to find woodland megafauna. Hawthorn can easily be recognized by a combination of traits: they are the only shrubs

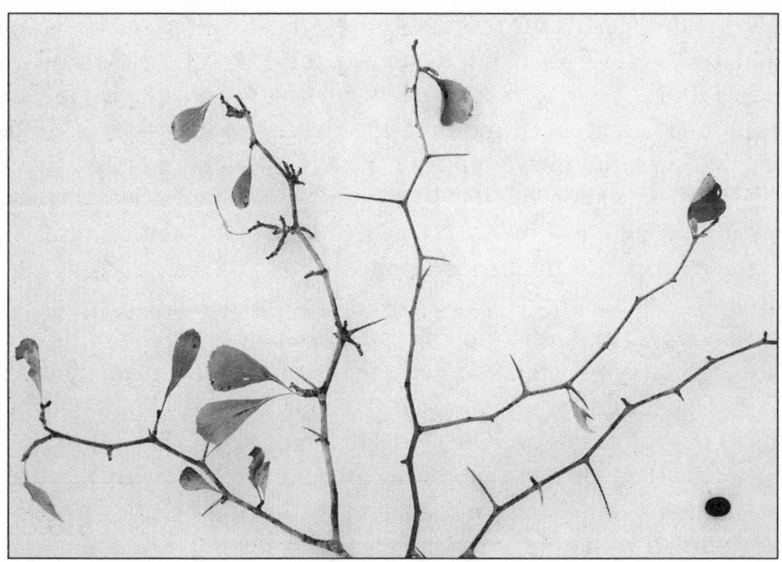

ARMAMENTS ADAPTED FOR LARGE MUZZLES. This branch of hawthorn (genus *Crataegus*) was collected in Florida in December, after most of its leaves had fallen.

and small trees with long, straight thorns on tangled, zigzag branches that bear single leaves with toothed edges. The range of genus *Crataegus* extends beyond the eastern United States to woodlands of the western states and to wooded temperate landscapes of Europe and Asia. Hawthorn thrives in British hedgerows.

Dan Janzen judges that the small, applelike fruits of hawthorn are to some degree anachronistic.[21] That may be so, although I have watched robins consume the berries of ornamental hawthorn trees a block from my New York apartment. What should interest us here, however, are the widely spaced and menacing thorns. Hawthorn looks more like an African plant sculpted by an arms race with rhinos and kudus than a companion of America's white-tailed deer.

A plant has two choices for defending its leaves against herbivores. It can deploy a physical defense—with thorns or prickles on branches, spines on leaves, or an entangled growth habit—or it can defend its leaves chemically. Hawthorn has chosen physical weaponry that, given today's challenges, overshoots the mark. The thorns are too long and insufficiently dense. Narrow-muzzled deer easily pluck around them.

R. F. Mueller regards the thorns as anachronistic for exactly this reason. Ground sloths and mastodons are among the ghosts he sees in the vicinity of hawthorn. "The browsing habits of these larger animals must have been quite different from those of the delicate-muzzled deer. It seems likely that they would have broken off whole branches or engulfed entire branch ends, stripping off the leaves and small twigs, as existing elephants do. Feeding leaf by leaf as deer and goats do would not have satisfied their enormous intake requirements."[22]

Susan M. Cooper and Norman Owen-Smith studied the feeding styles and preferences of kudu, impala, and Boer goat in Africa. All three species are adapted for browsing thorny plants, but they feed in different ways. The goat and impala pluck leaves one by one, carefully avoiding thorns. The kudu, in contrast, "bit off large shoot ends between their cheek teeth, then turned the shoots around in their mouths until the spiky leaf tips pointed outwards before swallowing." Only when thorniness is coupled with especially small leaves is the kudu turned away. The cost of contending with thorns is not worth a paltry gain. During severe droughts, however, the only foods available are those with thorns and small or sparse leaves. In such times the throats and esophages of kudus killed by hunters "showed a profusion of scar tissue." The authors conclude that "the main effect of thorns and spines is to restrict the bite sizes that browsing ungulates obtain, in effect increasing the handling time per unit of food ingested." African plants also defend against megaherbivores by producing thickets. Thickets and thorns are troublesome for browsers, especially if compounded in the same plant, as in hawthorn.

One final trait of interest is the height above ground that thorns disappear. Small thorny trees, like hawthorn and mesquite, may bear thorns well above the reach of deer. Is this miscorrelation of thorniness and browsing height meaningful? Cooper and Owen-Smith point to European holly *(Ilex aquifolium)* as an example of a tree that bears armaments—in this case, leaves with spiny edges—at a height inordinately far above the reach of today's browsers. Only at four or five meters do the spines disappear.

While working on this chapter, I visited a hardwood forest in the panhandle of Florida that contains American holly *(Ilex opaca)* as well as hawthorn. The distinction between spiny and smooth-edged leaves on the holly was easy to spot on a leaf-by-leaf basis. So, too, was a range of thorniness among the hawthorns—within a single plant as well as between plants. I was wary, however, of judging correspondences between armament and altitude. It would be all too easy to see what I preferred to see.

Variations in armament among holly leaves. The European holly (upper) and American holly (lower) both produce spiny and smooth-margin leaves on the same individual. Spiny leaves predominate within three or four meters of the ground; smooth-edged leaves grow in the higher branches. The red berries of hollies have evolved to attract birds.

The matter is complicated by the fact that browsing can induce a scantily protected plant to produce defenses in profusion.

Several months later I observed European and American holly side by side in the Brooklyn Botanical Garden. Both species carried their spines to a height of perhaps four meters—far too high to be threatened by deer. Is this deployment of spiny leaves by tall hollies a vestige from the era of giants?

Botanist Peter White has studied the correspondence of physical defense with height for devil's walking stick, *Aralia spinosa* (family Araliaceae).[23] Inspired in part by Janzen and Martin's anachronism idea, White decided to rigorously test his casual observation that the growth pattern of thorns (technically, prickles) on devil's walking stick appears to be better matched with the missing megafauna of American woodlands than with browsers alive today.[24] Devil's walking stick, true to its name, is more a tall stick than an arborescent tree. The prickly stem may grow three meters before sending forth any branches at all. It is extremely vulnerable to browsing before branching occurs, because loss of the terminal bud is a serious setback in its race to rise above browsing range. The plant must

DEVIL'S WALKING STICK. A young *Aralia spinosa* is armed with prickles. It will continue to produce prickles far above the height of today's browsers in eastern American woodlands. Is the plant still defending against mastodons?

then resprout from the side of the lone stem, renewing its skyward climb. Like hawthorn, devil's walking stick favors sunny, disturbed patches and margins of the deciduous forests of eastern North America.

Peter White undertook his *Aralia* study in the Great Smoky Mountains of Tennessee, where he measured the distribution and size of prickles as a function of age and height in a sample of more than a hundred plants. White concluded that "if prickle distribution reflects the vulnerability of *Aralia* to browsers, then *Aralia* stems are most vulnerable when young and within three meters of the ground." He speculated that *Aralia* prickles might be anachronistic—well matched for the big herbivores of the Pleistocene but overbuilt for the tallest browsers today.

The same excursion in northern Florida that introduced me to hawthorn and holly trees also offered a chance to examine *Aralia*. The place was Torreya State Park. I was there to witness the world's most endangered gymnosperm tree, *Torreya taxifolia,* in the sole pocket of native habitat that remains. More is said about this tree in Chapter 10, as anachronistic fruit may have contributed to its decline. The point here is that anachronisms are not hard to find. In one small reserve in Florida, I found a tree with anachronistic fruits, another with anachronistic thorns, a woody stalk with overbuilt prickles, and a fourth plant with spiny foliage at a height that cannot easily be explained by association with living fauna. Looking for one candidate anachronism, I was surprised by three more. If one has learned to spot them, a walk in the woods or a trek in the desert will almost surely offer glimpses of ghosts.

The year that Dan Janzen got his paper with Paul Martin into print, he also had an essay on anachronisms published in the magazine of the Missouri Botanical Garden. He wrote, "Botanists have long been unable to explain why many New World woody plants are spiny or thorny, even in areas where plant-eating animals are absent. Honey locust, acacia and mesquite in the New World produce a heavy skirt of thorny branches in the lower part of the crown, and nearly thornless branches in the upper part. . . . The large mammals we lost in the Pleistocene were not only fruit-eaters and grazers; many of them were browsers. It is not hard to imagine that spines or thorns are remnants of the constellation of traits that once protected woody plants from excessive browsing but haven't been needed in the last ten thousand years, a mere moment in evolutionary time."[25]

Notice that Janzen includes honey locust in the list of thorny trees, along with acacia and mesquite. Fruits are thus not the only anachronistic feature of this legume of eastern woodlands and savannas. Honey locust is, in fact, the queen of thorns in North America. Its thorns are exceptionally long, multipronged daggers. The species designation of the taxonomic name, *Gleditsia triacanthos,* is Latin for "three-thorned." All thirteen members of genus *Gleditsia,* most of which are Eurasian, are thorny. The Argentinean species, *G. amorphoides,* bears the most outlandish thorns of all—as long as forty centimeters. Regionally it is known as *espina de corona Cristi,* Christ's crown of thorns.

Whatever the species, *Gleditsia* thorns achieve their apex of glory on the trunks of younger trees. As the tree ages, its bark thickens and hardens, and the trunk thorns may subside—unless the bark is scraped or other-

THE QUEEN OF THORNS IN NORTH AMERICA. Honey locust bears thorns on the trunks of young trees as well as lower branches. On branches higher than about five meters, the tree ceases to produce thorns.

wise injured. Pruning (or browsing) will enhance thorniness on branches too. Thorns on the trunk of honey locust are probably intended to deter bark-stripping animals, such as hungry horses in the winter. The thorns are no match for elephant tusks, of course. But how then would an elephant avoid injury when eating the bark?

Another tree of temperate North America that bears anachronistic thorns is osage orange. "God designed Osage orange especially for the purpose of fencing the prairies," wrote Jonathan B. Turner in 1848.[26] Farmers tolerated the nuisance of big, inedible fruits in order to gain the services of the stems. Rows of planted seeds would transform into stems that could be pruned and woven into hedges "horse high, bull strong, and pig tight," thanks to a synergy of suckers and thorns.

Osage orange (renamed *hedge apple* in the prairie states) was a boon to agriculture. Farmers from Colorado to Illinois imported seeds from Oklahoma and Arkansas so that living fences could be nurtured around croplands. For these early settlers "a plow without a fence was a hammer without a nail, a rifle without a cartridge."[27] Thousands of miles of hedges were planted and pruned throughout the American prairie during the third quarter of the nineteenth century. The enthusiasm for hedge apple subsided only when barbed wire claimed the market for farm fences.

Plants Who Remember the Moa

Though formidable defenses against the muzzles of mammals, thorns are ineffective against the beaks of birds. Only one large, isolated landmass remained barren of soft muzzles until seafaring humans arrived with livestock. Only there did birds come to exclusively occupy the niche of megafaunal browser. And only there did a diversity of plants evolve a distinctive set of traits that deter megafaunal browsers by means other than thorns.

The land is New Zealand, the North and South Islands of which have a combined area as great as the British Isles or the state of Colorado. The megafaunal browsers are moas—ostrichlike birds driven into extinction less than a thousand years ago. The plants in question are fifty-four species native to New Zealand that share a thicketlike, divaricate pattern of growth.

Divaricate plants are not unique to New Zealand. They can be found in California chaparral and scrublands of Argentina and Madagascar.[28] Any plant producing stems that diverge at wide angles from the elder stem and that sprout abundantly from lateral buds as well as at the apex will tend toward thicketlike growth. If the apical buds are pruned by tooth, beak, or gale-force winds, the lateral buds will dominate the growth pattern, yielding a maze of interlaced stems. In lands outside New Zealand, many of these thicket-forming plants also defend themselves with thorns (hawthorn and barberry are examples). Almost all such outliers, however, lack the extraordinary combination of leaf and growth characteristics that make New Zealand's divaricates distinctive.

Those of us who live in landscaped neighborhoods have encountered plants that exhibit at least a modicum of divarication. Shrubby plants used for hedges and especially for topiary have dense growth that is enhanced rather than harmed by intense pruning. These cultivars derive from wild stock that evolved in places where the photosynthetic costs of this architectural design (a high degree of self-shading and a wasteful branch-to-leaf ratio) were more than matched by the benefits gained in deterring browsers or in standing up to the physical stresses of a windy beach or mountain ridge.

North America's yaupon holly, *Ilex vomitoria*, is among the best. It is a thornless and intensely divaricating shrub. Landscaping cultivars are derived from wild stock that grows on the Atlantic and Gulf Coasts, from

A DIVARICATE PLANT ADAPTED TO THE SEACOAST. TOP: Wild yaupon holly, *Ilex vomitoria,* has been sculpted by wind and salt spray along a Florida beach. BOTTOM: Close-up of the stressed, windward side of the shrub.

New Jersey south to Florida and west to Texas. In the wild, this holly fulfills a vital ecological function in stabilizing coastal dunes. So important is yaupon for beach stabilization that strong measures are now taken to protect the remaining clumps. Yaupon is also drought adapted, growing at the peaks of sand dunes and subjected to salt spray. Storms sculpt these exposed shrubs into dense, rounded cushions. Smaller dunes, in fact, may be no wider than the sprawling breadth of a single, ancient yaupon plant that shelters the sands. When the plant dies, the dune will vanish. A hundred meters inland and sheltered from wind, the same genetic stock produces much rangier but still strongly divaricate forms.

Yaupon is evergreen like the American holly and the familiar hollies of Christmas decorations, but the leaves of yaupon are small and smooth-edged rather than prickly. Easy to chew and blandly tasty, they would not stand out in a tossed salad. A completely uninformed guess would attribute yaupon's divaricate growth habit to both biological (browsing) and physical stresses. But look again at the species name: *vomitoria.* Native Americans made a drink from these leaves for ritual cleansing; in other words, to induce vomiting. Are the leaves also toxic to deer?

That question will remain unanswered, as our story is not about divaricates in America but about those in New Zealand. The most extreme of the divaricating plants native to New Zealand may, unlike yaupon, be ill suited to the aesthetic demands of suburban home owners. The reason is a paucity of greenery. New Zealand divaricates are densely branched but sparsely leaved, and those leaves may be exceptionally small, with the smallest borne on outer branches. Some of these divaricates, moreover, produce new growth which, instead of greening the plant, enhances the dominant browns and grays; in these plants, new growth mimics dead leaves.

Is this coupling of a divaricate stem structure with sparse, small, and sometimes cryptic leaves just the luck of the draw for New Zealand? Or does the suite of traits indicate adaptation? There is no disagreement that an adaptive explanation should be sought. The 54 species of divaricating plants with sparse and small leaves are not descendants of a single common ancestor that happened to suffer stem growth and leaf mutations in the distant past and then passed the traits on to a multitude of descendants. Rather, the 54 species are drawn from 20 distinct genera, representing 16 families of angiosperms and one family of gymnosperm. The divaricate suite of traits unquestionably owes to parallel, independent evolution in at least 17 (and more likely close to 50) separate lineages. What was driving that evolution?

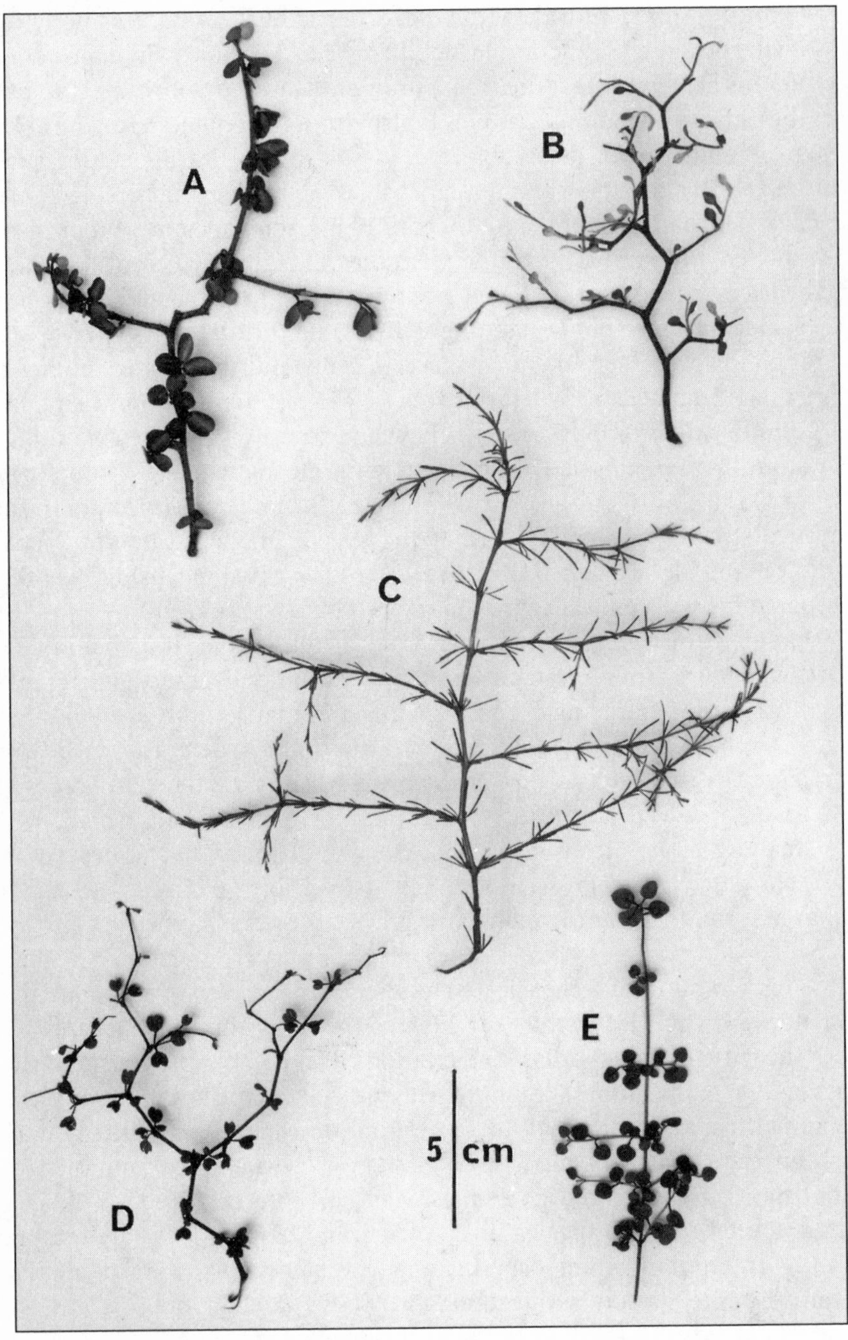

Plants moas left behind. These eleven species of divaricate plants native to New Zealand evolved within ten distinct taxonomic families. Courtesy of John Dawson from his 1993 book, *Forest Vines to Snow Tussocks*.

F
G
H
I
J
K

Divaricate plants are not occasional curiosities in New Zealand. One out of every ten woody plants native to the North and South Islands displays this growth form. Many are conspicuous and important members of their biological communities.[29] No place in the world is richer in divaricate growth.[30] Some unique or uniquely enhanced selective force must have strongly favored the divaricate habit in New Zealand.

Apart from the late Pleistocene, the post-Cretaceous fossil record of plants is poor in New Zealand, and the odd leaf or stem segment tells us little about architectural form. It therefore cannot be known when divaricates first appeared and when they became abundant. No theory of divaricate forcing will be able to tie into a paleontological record. Nevertheless, other clues can be discerned.

The first adaptive explanation that anyone offered placed the stimulus of the divaricate growth habit in the physical realm, supplemented by geographical and biological constraints.[31] The climate changed, it was argued. Some proponents of the climate theory speculated that divarication was selected during the glacial episodes of the Pleistocene. Others believed the suite of cool- and dry-adapted traits came into being much earlier during the Miocene, about twenty million years ago, when New Zealand's magnificent Southern Alps were thrust into existence. Until then, the islands had little topographic relief. Rain shadows—and thus aridity—would have developed leeward of the peaks, and cooler temperatures toward the peaks. Glacial changes and mountain building were not unique to New Zealand, of course, so climate change by itself cannot explain the abundance of divaricate species on the islands. That is where geographical and biological constraints come into play. Both owe to New Zealand's isolation.

Plants already adapted to cool and dry conditions could not be recruited from elsewhere when the need arose in New Zealand, theorists contend. Too much ocean separated New Zealand from everywhere else. Isolation thus posed a severe geographical constraint. New Zealand had to freshly evolve suitable plants from its own moisture-loving, subtropical stocks. There lay the biological constraint. The divaricate suite of traits may have been the easiest way for evolution to accommodate the change in climate. Thicketlike growth protects interior leaves from light frosts. The growth form may then become a heat trap on cool sunny days, thus invigorating the biochemical reactions upon which photosynthesis depends. Interior leaves are also less exposed to desiccating winds.

Today's advocates of the climate hypothesis still point to the frost-screening, heat-trapping, and wind-breaking advantages of divaricate plant growth. They have, however, retooled the supporting rationale in a way that better meets the demand for an explanation unique to New Zealand.[32] It was indeed a cooling and drying that pressured subtropical lineages to assume divaricate forms, they suggest. Yet just as important was that episodes of frost and drought were (and remain) untethered to the calendar.

Continental plants that must survive deep winters may drop their leaves at the beginning of hard times. The deciduous tactic is also apparent in warm regions with distinct and long-lasting dry seasons. In New Zealand, however, the climate is best described as variable. New Zealand plants must cope with a "high and unpredictable incidence of frost, wind, and drought."[33] Stressful weather can happen at any time of the year but usually does not last long. Conditions favorable to photosynthesis rapidly resume. A sacrifice of leaves makes sense if cold or drought can be expected to continue for several months. But if temperatures dip below freezing for a few days one month and maybe a few days the next, a better strategy is to bulk up the leaves to withstand cold snaps and to redesign the branching architecture so that much of the foliage will be spared frost damage. This is what New Zealand's divaricate plants have opted to do, the climate proponents maintain. New Zealand is overwhelmingly a land of evergreen plants, and most divaricate plants are, in fact, evergreen. The proponents of the climate hypothesis argue that a protective cloak of interlaced branches would be especially helpful for plants that grow in the open or along the edges of forests where the chill of night is greatest. Wind-whipped coastal habitats and mountain ridges would thus not be the only environments that would nurture the divaricate habit.

Divaricate plants in New Zealand do predominantly live under open skies. They tend to be forest-edge or scrubland plants.[34] That fact, however, does not support the climate hypothesis to the exclusion of a competing theory. It is equally conducive to an entirely biological explanation: big browsers. Forest edge and scrubland are precisely the habitats big browsers frequent. In the depths of forest, foliage is out of reach; in the grasslands, grasses (and thus grazers) take over. The intermediate zones are where large browsers prosper and exert selective pressure. Spiny plants of the Americas (cactus, mesquite, hawthorn, honey locust) are found primarily in such habitats. One might expect the same of divaricate plants in a land of browsing beaks.

The only large browsers thought to have lived in New Zealand since the Cretaceous were the flightless moas. If divaricate plants were shaped by a biological interaction rather than by climate, moas are the only conceivable candidates.

The world's biggest flightless birds of today (or only recently sent into extinction) are all thought to share a common ancestor that lived during the Mesozoic on the supercontinent of Gondwanaland, which was an immense aggregation of all the landmasses now strewn across the Southern Hemisphere (New Zealand included). When Gondwanaland ruptured in the Cretaceous, these ratite birds continued their evolution independently on each of the fragments. Over time, ancestral ratite stock evolved into the ostriches of Africa, the rheas of South America, the emus and cassowaries of Australia, the giant elephant birds (recently extinct) of Madagascar, and the moas of New Zealand.

Ostriches share the African savanna with a great number of mammalian herbivores of placental stock. Rheas, until the end of the Pleistocene, shared the South American pampas with large placental mammals too. Not long ago emus cohabited the Australian grasslands and scrublands with great marsupial herbivores, while the fruit-eating cassowaries kept to the forest. Twice the size of an emu, *Genyornis* flightless birds (not ratites, but closer in ancestry to ducks) also inhabited Australia when humans first arrived. Madagascar had, until we overhunted them, large herbivorous mammals—giant lemurs—with whom the elephant birds unique to the island must have shared the browsing niche. Only in New Zealand did the great flightless birds have no mammalian competitors. Only in New Zealand were there no broad, soft muzzles shaping the evolution of plants.

Eleven species of moa inhabited New Zealand during the Pleistocene and well into the current Holocene epoch. All were rapidly driven to extinction by a seafaring Polynesian people who colonized the islands eight centuries ago. By the time Europeans arrived, even the most remote populations of forest-dwelling moas were gone. Subfossil evidence from kill and processing sites indicates massive exploitation. Paul Martin's overkill theory may be controversial for the mammoth extinction in the Americas thirteen thousand years ago, but it is virtually unopposed as an explanation for the demise of New Zealand moas.[35]

The smallest species of moa was the size of a turkey, the largest a bit taller than an ostrich and a great deal stouter. Examination of moa gizzard contents has led experts to conclude that the tallest species of moa were browsing specialists. Twigs neatly sheared into lengths of one to three centimeters compose the bulk of the vegetational mass, the remainder being leaves and seeds.[36] (Some smaller species of moa may have primarily grazed grasses, as do several kinds of flightless birds still alive in New Zealand today.) Because all birds lack teeth, moas depended on muscular gizzards equipped with swallowed stones to process high-fiber vegetation.

A ghost of New Zealand. The largest moas were taller than ostriches and considerably bulkier. Unlike ostriches, the moa lineage entirely devolved away wings. So did New Zealand's kiwis, depicted here foraging around a moa's feet. All dozen species of moa are gone.

The notion that an extinct race of flightless birds might best explain the abundance of divaricate plants on New Zealand was sporadically proposed beginning about forty years ago. It was not until 1977, however, that the idea was fully elaborated, and with arguments refuting the climate theory. This was accomplished by a pair of New Zealand biologists, R. M. Greenwood and I. A. E. Atkinson, in a paper published in the *Proceedings of the New Zealand Ecological Society*.

Greenwood and Atkinson pointed out that New Zealand's divaricate plants do not favor coastal and mountain habitats where adaptation to wind could be invoked. Neither are they prominent members of vegetational communities on exposed cliffs, where wingless bipedal vertebrates would surely have been excluded. Other evidence in favor of the moa theory is that the vegetational communities on lesser islands support fewer divaricate plants. New Zealand has myriad small islands offshore from the two big islands, and these lesser islands would not have supported moa populations.

One final thread of evidence is especially intriguing—and perhaps the least amenable to climate arguments. Nine of the fifty-four species of divaricate plants are trees rather than shrubs, yet they do not carry a thicket of growth all the way into the forest canopy. Instead, divaricate growth occurs only in juveniles. By the time one of these trees is three or four meters tall, its growth pattern switches to a mode that produces an arborescent architecture. The leaves undergo an equally dramatic change, enlarging and sometimes adopting an altogether different shape. Trees with divaricate juvenile stages seem to be found elsewhere only in Australia, and there the species number but two (in family Rutaceae).[37]

Overall, the moa theory seems to have garnered the most converts, despite having "a science fiction ring to it."[38] Even the climate advocates concede that they are in the minority.[39] Few of the involved scientists, however, favor the moa explanation to the exclusion of climate. In their 1977 paper that launched the moa theory, Greenwood and Atkinson proposed that moas were "primarily" responsible for this growth form on New Zealand. Later, Atkinson and three coauthors concluded it was "probable" that "climatic and browsing factors acted in synergy."[40]

It is reasonable to conclude that the suite of branching and foliage traits common to New Zealand's divaricate flora is to a great degree anachronistic. Ten percent of the indigenous woody flora are still visibly affected by the ghostly presence of extinct browsing birds. Might moas have left behind other anachronistic features of plants? Might some New Zealand fruits have forged mutualistic relationships with the giant birds? Greenwood and Atkinson mentioned this possibility in their 1977 paper but chose to limit their work to vegetative anachronisms (the list of which they later expanded to include tussock plants prevalent in New Zealand grasslands).[41] The fruits of New Zealand's divaricate plants are, however, too small to be megafaunal. And they tend to be too entwined within thickets to provide easy plucking. Even small flighted birds cannot work the interior of the densest divaricates. Stone-equipped gizzards would have thwarted any lineage that attempted to recruit the moa digestive tract for seed dispersal.

The fruits of New Zealand's divaricate plants, while not anachronistic today, may become so in the near future. Most are adapted for dispersal by lizards, which can climb into and through the interwoven stems. Lizard populations are severely threatened by rats, cats, and weasels (stoats) that have been accidentally or intentionally introduced to the islands. New Zealand is, in fact, a superb spot to experience the horror of human capacities to modify landscapes. There we can grasp the practical and moral consequences that await the rest of the world when our species fulfills its sad destiny of transforming a diverse Earth into a "planet of weeds."[42]

FECAL CLUES. Hindgut processors such as horses pass larger fragments in their dung (left) than do forestomach processors such as deer (upper) and elk (lower). Large-seeded plants may therefore require hindgut processors for dispersal. Honey locust seed (1 cm) for scale.

8

Who Are the Ghosts?

"FRANKLY, THIS IS NOT REALLY SCIENCE. You haven't got a way of testing any of this. It's more metaphysics." Richard Tedford is a vertebrate paleontologist at the American Museum of Natural History in New York City, where he curates the immense collection of mammal fossils from the Cenozoic. He voices these doubts as I sit in his office, beginning to pull fruits out of my bag.

"That may be true," I say, rummaging inside the backpack. "But we might be able to broadly identify which sorts of animals a fruit may have been partnered with and which ones probably not. We know there are things here today that are not adapted for these times—we can say that much. We can't say for sure an avocado is still waiting around for a gomphothere to arrive, but it's waiting for something."

"Gomphotheres are late arrivals in North America," Tedford reminds me. "So the relationship has to long predate them."

"Right. So what about brontotheres or maybe an old rhino?"

"Most of the early brontotheres and rhinos had very low-crowned teeth," he responds. "They would have husked and eaten the flesh, but whether they would have cracked the nut, which is what you are really after . . ."

"Oh no, no. I don't want them to do that," I exclaim.

It doesn't take long for Tedford to catch on to the anachronism idea and to heartily offer the assistance I am looking for. We agree that science or no, trying to find the most (and least) likely dental matches for one anachronistic fruit or another is great fun. We eventually agree that, for my purposes, the Pleistocene menagerie is all that need be considered. A fruit form may have been shaped initially by brontotheres during the early Cenozoic and maintained by rhinos during the mid-Cenozoic, but the megafauna of the Pleistocene would have carried on the mutualistic tradition until just thirteen thousand years ago.

"Are you familiar with this?" I ask.

"No."

"They're called osage oranges, genus *Maclura,* part of the Moraceae family. They have a white, sticky latex inside when they first fall. They are very similar to *Treculia africana* in Africa that the elephants and rhinos love. But it's poisonous to people. Osage oranges used to be all over North America in the Pleistocene, and a similar genus in Asia is *Cudrania.* After the Pleistocene, by the time the Europeans got here, osage orange had constricted down to just a little bit of northern Texas. So the thought is that nobody was dispersing it. And nobody is dispersing it today, except humans.[1] After the latex dries squirrels do go after the seeds, but no native animal is eating all this pulp."

"So the seed is not going intact through a gut today," he surmises.

"Right. Osage orange is one of my key examples. I'm calling it an extreme anachronism."

By now I have handed him the fruit. Tedford pauses while examining it, then continues, "So what you are thinking is that this thing would have been chewed up and swallowed by something, without disturbing the seeds, or at least without disturbing many of them. This has got a lot of seeds in it?"

"Yes, lots. They're about the size of orange seeds."

"And it's got a toxin in it," he says. "It's got its own defense."

"I think the toxins are primarily against insects. But mammals have a variety of ways to counteract such things." I recite several examples, then ask, "So what do you think?"

Tedford rolls the fruit in his palm. "Well, it's a pretty ugly-looking thing," he begins. This is February and the once lovely green fruit is marred by patches of brown, despite my attempts to preserve it by refrigeration and a varnish of clear resin.

"Osage orange," he says slowly, contemplating the fruit. He pauses then says, "Proboscidea should be a good match. Both the elephantine genera and the short-jawed gomphotheres would have been available during the Pleistocene."

We move on to the next fruit, that of Kentucky coffee tree, but come to no conclusion. Then I hand him desert gourd.

"These guys have a tough rind," I begin. "Just like a squash, they've got nice pulp on the inside." I tell him about Dan Janzen's experience with another smooth, spherical, tough-husked fruit in Costa Rica, *Crescentia,* and about how Janzen observed that cattle cannot break through the rinds. Horses can—but the biggest fruits stymie even their dental equipment. "It would seem that something had to have incisors to eat desert gourd," I suggest.

REUNITED: MACLURA AND MASTODON. Shown here is surely the first time in thirteen thousand years that the fruit of osage orange, *Maclura pomifera,* has touched a molar of its missing partner in evolution, *Mammut americanum.*

"Both upper and lower incisors," he clarifies, as cattle have lower incisors but lack uppers.

I have visited the two great halls of extinct mammals many times, yet it is sobering how little one sees without precise intent. A week before my meeting with Richard Tedford, teeth were my quarry. I tried to ignore the majesty and strangeness of the bones to concentrate on teeth. The glyptodont skeleton drew my attention more than any other. A relative of the armadillo, it resembles a great tortoise or ankylosaurid dinosaur more than a mammal. The one on display is of medium build, but its bony "shell" is nevertheless enormous. The largest glyptodonts were three meters long and weighed a ton; Paul Martin likens them to Volkswagen beetles. Nose to the ground, bulky, and surely restricted to level terrain, the glyptodont would have overlapped ecologically with the gourds of disturbed soils in full-sun landscapes. Would its oral equipment have been up to the task?

Glyptodonts had flat-topped, ever-growing molars, but nothing else—no teeth up front. Tedford did not regard glyptodonts as a good match for the gourd, although there would have been plenty of other matches: all the proboscideans and horses, plus toxodon in South and Central America, and a late lineage of camel that, unlike the living camelids and cattle and deer, had not yet traded away oppositional incisors for an upper horny plate. The camels, as well, kept their canine teeth, which might have been useful in cracking gourds. Fundamentally, Tedford didn't think that the glyptodonts would have had sufficient gape. Getting a gourd into a mouth is only the first step. One then has to be able to manipulate the object against a cutting surface. I am surely not alone in having more than once been bested by a cherry tomato. A cherry tomato easily pops into the mouth but may be too big to fit between molars. You can't chew, you can't talk, and if you are not on intimate terms with your dinner companion, you can't easily resolve the impasse.

No, Tedford told me, desert gourd is probably not to be paired with glyptodonts. Proboscideans, yes; horses, yes; toxodon, yes; but glyptodonts, no. He did offer another candidate mutualist that is, in a way, more delightful. Through the late Pleistocene, one reptile counted among the megafauna of North America. This was a giant tortoise, somewhat smaller than its living kin of the Galápagos Islands. *Geochelone* is the genus of both, and North America was home to several species during the Pleistocene. Not long ago, giant tortoises lived in warm landscapes the world over. Because they easily float and can survive for weeks, even months, without food or fresh water, giant tortoises colonized tropical lands east and west. But they cannot survive the human presence. Giant tortoises vanished wherever and whenever humans arrive.

Only on the most remote islands did giant tortoises remain into the modern era. Their extraordinary ability to survive deprivation, however, became a liability. European sailors began to visit tortoise islands expressly to fill ship larders with live animals stored helplessly on their backs. Crews could then enjoy fresh meat for months to come. Today giant tortoises are found on just a few small islands far removed from the closest landmass. The Galápagos Islands off Ecuador and the Aldabra Islands north of Madagascar are so remote and so steeply cliffed that the press of human exploitation was delayed long enough for the conservation-minded to intervene. For the Paleo-Indians who first arrived in North America, chancing upon *Geochelone* guaranteed a feast. The tortoise would have been easy to spot, easy to overtake, and easy to kill.

Hindgut Heaven

An awareness of ecological anachronisms ignites a fascination for more than strange fruit. One wonders about the missing partners, the ghosts of evolution. To date, researchers have been content to link anachronistic fruits in the Americas with the broad guild of extinct Pleistocene megafauna. We know the beasts were biggish, and we know they are gone. But is that all we can say?

When standing in the kitchen slicing an avocado, I want to know whether a gomphothere, a ground sloth, a notoungulate, or perhaps a mixed-species group is what I hear rumbling in the living room. When passing a honey locust on Laguardia Street, I want to know who exactly the pods are waiting for. Did camels or elephants have first dibs? Should I perhaps listen for the approach of horses?

To answer these questions we must look at the animals from the fruit's point of view. To begin, we need to study the teeth, as Richard Tedford and I have already done in a cursory and lighthearted way. Teeth may be a help or a hindrance to seed dispersal. Front teeth may be required for cracking the rind of a gourd, but too much time spent in the company of molars may turn the next generation to mush. On the other hand, some contact with a grinding surface may be necessary to liberate seeds from a woody pod. If evolution has fostered a seed that can survive the destruction of its pod, then such abuse may now be indispensable for germination. Without scarification of the tough coat, honey locust and gymnocladus seeds cannot absorb the water they need to sprout.

For one category of megafaunal dispersal agents, the dental challenge may not end when a seed is swallowed. All but the smallest seeds are destined to meet the molars again when food fermented by microbes in the forestomach is returned to the mouth for another round of mastication. Most members of order Artiodactyla, the even-toed ungulates, chew their cud. Deer, elk, cattle, sheep, goats, giraffes, antelope, pronghorn, and camels all remasticate their food. Hippos and peccaries also ferment their food in forestomach compartments before subjecting the mass to gastric juices in the final, enzymatic section, but they do not regurgitate and rechew the contents.

While a deer or cow or camel is chewing a fibrous clod of cud, a seed within (perhaps softened by soaking in the forestomach) may slip past the molars a second time or else it might be crushed or ejected.[2] Seed ejection is not helpful from a plant's perspective when the fruit is first encountered under the parent tree. On the second time around, however, seed spitting may or may not constitute effective dispersal. It all depends on where an animal chooses to chew its cud. If a bovid is resting in the shade of a favorite tree out on the savanna, the seed's future may not be bright. Similarly, if a flock of parrots is tracking a herd of cattle, the rumen ejecta may be scavenged by the birds, whose powerful beaks crack the pits or seed coats and steal the contents.[3] Or perhaps mice will scavenge spat seeds.[4]

There is no second time around for hindgut processors. Food transits from the gullet immediately into a stomach that secretes digestive juices. Because hindgut animals have only one opportunity to physically abrade

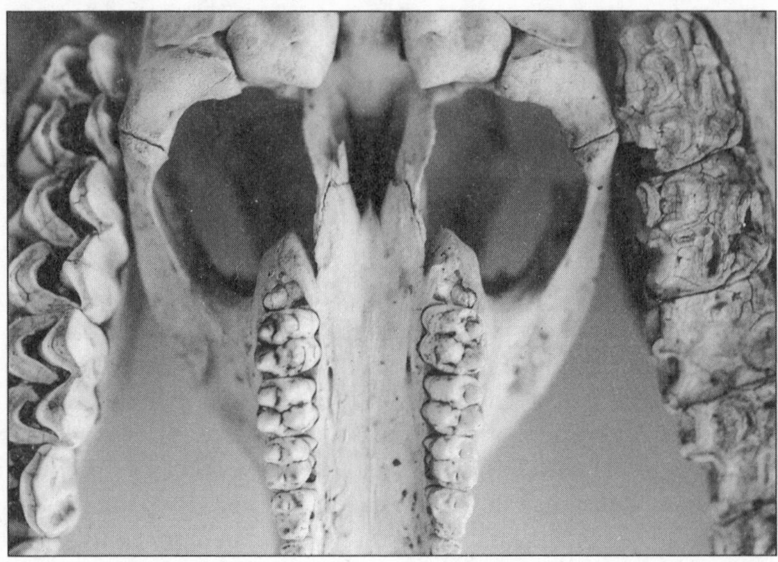

MOLARS OF BROWSING, SEED-CRUNCHING, AND GRAZING HERBIVORES. LEFT: Elk, adapted for browse, forbs, and tender leaves of grass. MIDDLE: Peccary, adapted for roots, grubs, and seed predation. RIGHT: Horse, adapted for grazing silica-studded stems of grass; note the severe abrasion on the horse teeth.

fibrous material, they tend to chew each mouthful more thoroughly than would a ruminant of the same size. Thus, they may at the outset liberate (and eject) more seeds from pulp and fibers.[5] Context, too, is important. Dan Janzen observed that the degree of seed spitting in horses consuming guanacaste pods in Costa Rica depends on whether the animals are competing with peers for a limited supply.[6] How fast we eat determines how carefully we chew.

Who are and were the hindgut processors? The microbial vats within elephants, tapirs, rhinos, and horses all reside in the large intestine. A massive colon with undulating walls is the primary site. A long and elaborate side pouch (a cecum) at the upstream end of the large intestine provides supplementary fermentation. The extinct kin of today's hindgut processors—among them, gomphotheres, mastodons, and mammoths—surely were hindgut processors too. The coarse texture of mummified dung would seem to confirm the conjecture. Because the tree sloths of South America have the largest and most complex forestomachs of any animal alive today, one would suspect that extinct ground sloths would have relied on forestomach processing.

From a fruit's perspective, are there important differences between forestomach and hindgut mammals? Some forestomach fermenters cannot offer big seeds a nursery of dung. Deer, cattle, and other ruminants have only a small orifice separating forestomach compartments, and that orifice may prohibit even midsize seeds from passing.[7] Call to mind the images of a horse turd *(Equus)* and a cow pie *(Bos)*. How do they differ? Setting aside the fact that the first is much drier than the second, the main difference is that, even if grazing the same pasture, a horse will defecate larger plant fragments than will a cow. A horse does not digest its food as thoroughly. An even better comparison is shown in the photograph that opened this chapter. Here you can see the differences between the turd of a horse and that of two ruminants who also defecate dry, well-formed dung: elk *(Cervus)* and deer *(Odocoileus)*. Moose *(Alces)* pellets would have offered an even better comparison, because this ruminant can be as big as a horse, yet the dung is only twice the size of an elk pellet and of the same fine-grain texture.

A horse must consume twice as much fibrous food as a ruminant of the same size in order to extract equal energy. Hindgut megafauna are bulk processors to an extreme. The more fibrous the fodder, the more they eat; the more they eat, the faster the ingested material exits the other end. Some intestinal contents may be diverted into the cecum for more extensive processing. Because the cecum is a side pouch, however, other materials continue to pass unimpeded through the colon, which is the primary chamber for microbial fermentation.

Hindgut processors prefer quality, but they can make do with quantity. Ruminants, in contrast, have an anatomy that excels only when they can pick and choose the most digestible forage. Cell contents and the cellulose of plant walls is good; the lignin of woody stems is bad. A cow's forestomach becomes a liability on seriously degraded range, as it will have to eat *less* as the food value of forage declines. Why is this so?

Rather than cream-skimming each meal for the cell contents and the plant fibers easiest to break down, a ruminant ruminates until every piece has been fragmented to bits. Physiologically it has no choice. The orifice that allows one fermentation chamber in the forestomach to empty into the next is far smaller than gullet or anus. A plant fragment will not pass until it can be squeezed through that orifice by muscle contractions. This constraint applies to seeds as well as to stems. For ruminants, therefore, food must be chewed and rechewed until it can pass through the gate (technically, the reticulo-omasal orifice). The longer a cud chewer and its microbial symbionts take to satisfy this threshold, the longer between meals. If the forage is really lousy, the forestomach will clog and nothing will pass. Cattle can die of starvation while their rumens are bulging with food.[8] Horses (better yet, burros) on the same range will not be happy, but they will survive.

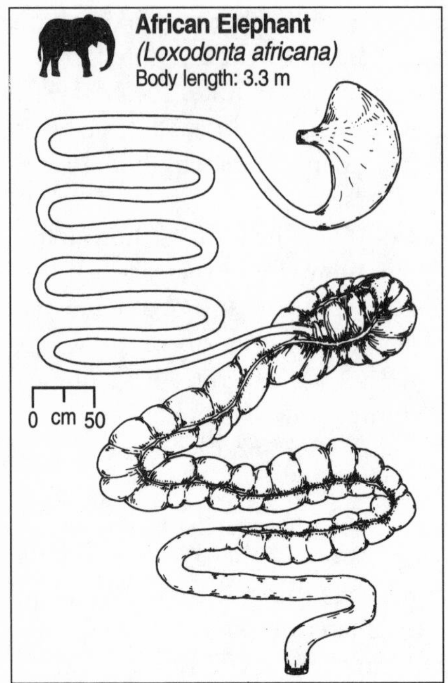

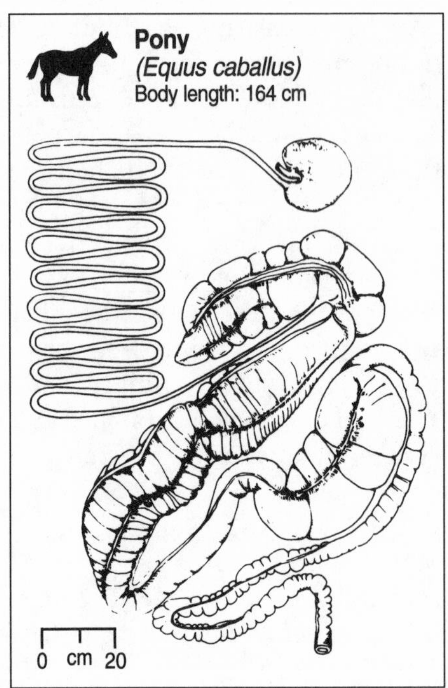

THE RANGE OF VERTEBRATE DIGESTIVE SYSTEMS. The elephant and pony are hindgut processors, whereas the llama provides for microbial fermentation in an enlarged and compartmentalized forestomach. The black bear, an omnivore, has inherited the simple digestive system of its carnivorous ancestors. Courtesy of C. Edward Stevens and Ian D. Hume from their 1995 book, *Comparative Physiology of the Vertebrate Digestive System.* Reprinted with the permission of Cambridge University Press. (Elephant anatomy after Clemens and Maloiy 1982.)

Imagine a bucolic landscape on a hot summer day. Horses and cows share the same pasture. How do we picture them? The horses will be heads-down grazing. The cows? Resting in the shade, ruminating.

Think, too, of deer, who may venture out of hiding only at dawn and dusk to browse. A horse, a rhino, an elephant would starve if pastured for so little time, especially where quality food is unavailable. Big animals cannot afford to be picky eaters. If you carry a lot of bulk, you must opt for bulk foods. An elephant doesn't pluck a leaf here and there. She rips off a branch or pulls up an entire clump of grass, roots and all. Asian and African elephants may spend more than eighteen hours a day actively foraging.

Both sets of herbivores, however, employ the same crew of microbial symbionts. The kinds of microbes found in the forestomach of a llama or a cow are exactly those found in the hindgut of an elephant or a horse.

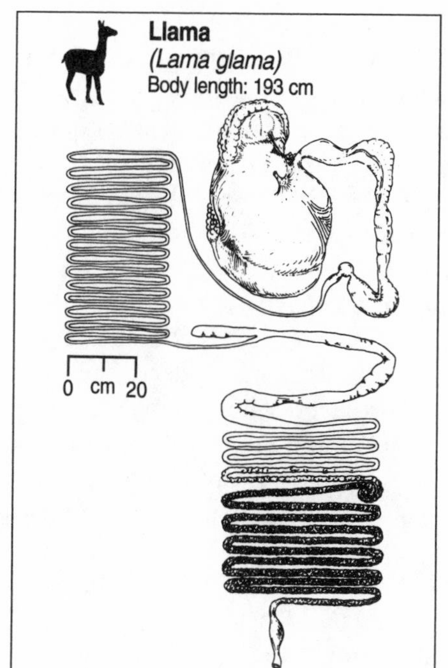

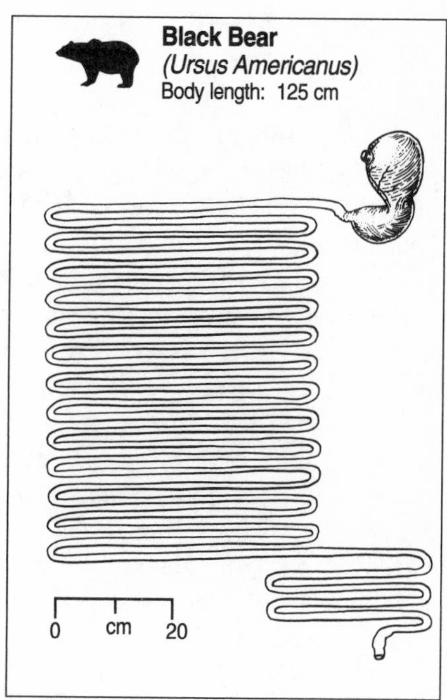

THE RANGE OF VERTEBRATE DIGESTIVE SYSTEMS *(continued).*

The symbiotic talent of gut microbes—fore or aft—is extracting food value from cellulose, which plants manufacture to support their cell walls. No animal, not even an insect, can produce an enzyme capable of extracting food value from cellulose. Mammals, birds, reptiles, and insects that consume fibrous plants all rely on microbial mutualists to do the work for them. No bacteria or protist that makes its home in a gut can, however, break down the complex lignin molecule found in woody plants. Cellulose, yes; lignin, no. Fungi are the only organisms up to the task, and the kinds of fungi that process lignin are not gut constituents. Even the American beaver *(Castor canadensis),* who eats the branches of willow and cottonwood trees, is not obtaining food value from the outer bark or inner wood—only from the thin green layer of cambium that resides in between.

Though equipped with the same microbial helpers, ruminants are less able than hindgut processors to move big chunks of food through their systems.[9] For example, white-tailed deer must spit the seeds of the Costa Rican tree *Spondias mombin*—if not while ingesting the fruits, then during rumination. Yet hindgut tapirs regularly swallow and defecate these seeds.[10] The Old World camel is a cud-chewing, forestomach processor that passes only small seeds from the fermenting forestomach to the enzymatic secretory stomach. No object wider than half a centimeter will make that passage.[11] Honey locust (with seeds 0.7 cm wide by 1.0 cm long) could not therefore have been enlisting the fecal services of Pleistocene camels in North America—that is, assuming the physiology of North America's *Camelops* bore as great a resemblance to today's Old World *Camelus* as its skeleton would indicate. For this reason, I should not conjure a camel on the streets of New York.

I cannot find information about the forestomach constraints of deer and elk in North America, but it is reasonable to assume that their forestomach orifices would be no larger than a camel's. Indeed, the seed of honey locust is almost as big as an entire deer pellet. Native deer do, however, defecate intact seeds of the tropical and subtropical American acacia tree, *Acacia farnesiana,* whose spherical seeds are about half the size of the ovate honey locust.[12] A variety of ruminants in southern Africa—giraffe, impala, and kudu—effectively disperse through their pellets the seeds of *Acacia* species native to that continent, including the picturesque umbrella thorn tree.

There can be little question that cattle brought to North America offer greater dispersal opportunities than do the native ruminants who survived the end-Pleistocene extinction. Dan Janzen found in the dung of cattle intact seeds from guanacaste pods, which are larger than those of honey locust.[13] No one, to my knowledge, has tested the fate of honey locust seeds on North America's biggest native ruminant: bison. What about the Western Hemisphere's other anachronistic fruits? Would seeds too big to pass through contemporary forestomachs have been blocked by the largest forestomachs present during the Pleistocene? If ground sloths did indeed house their microbes up front, from which end would avocado seeds have exited?

One other physiological constraint makes forestomach fermenters, as a class, inferior to hindgut animals as potential seed dispersers—at least for the most sugar-rich or acidic fruits, whatever the seed size. These animals cannot allow their microbial chambers to become acidic. Microbes that process cellulose demand a slightly alkaline environment, which the forestomach group fosters by producing slightly alkaline saliva—and being careful about what they eat. Deer have no interest in pawpaw for this reason. (How deer are occasionally able to gorge on apples is a mystery to me.) Colobine monkeys, who are forestomach processors of a very leafy diet, avoid not only acidic fruits but virtually all ripe fruits.[14]

If a forestomach acidifies, the good microbes—those that can digest cellulose—begin to die, and unhelpful or downright nasty microbes take over. If acidosis progresses to an extreme, lesions may develop and a frothy stew may impede belching, allowing gases to build. A sheep or cow can die of such "bloat" if gas pressure interferes with expansion of the lungs or beating of the heart. In contrast, a hindgut processor shunts acidic fruit directly into a stomach that thrives on acidity, producing its own acidic enzymes to break down energy-rich foods. By the time the easily digestible fare has been absorbed into the bloodstream through the lining of the small intestine, the remainder passes into the hindgut chambers at a pH favored by cellulose-digesting microbes.

Not only acidic foods, but sugar- and even starch-rich foods can foster acidosis in a forestomach. Bacteria effect the transformation, just as they do in converting fragments of candy in a human mouth into tooth-decaying acids. Cattle can develop acidosis even on a diet of grass if they feed too extensively on the sweet new growth of spring.[15]

More fundamentally, feeding a sugary or starchy fruit to a ruminant is a waste of food. Why? Animals can digest sugars and simple starches directly, without the help of microbes. That's why we experience a "sugar rush" from soda pop. A hindgut processor that consumes a sweet pawpaw, the sugar-rich pulp of a honey locust, or even the starch of a great ball of maclura will harvest all that energy for its own cells. Only cellulose, pectin, and other fibrous materials in the fruits will pass on to the cecal or colonic microbes. In contrast, a ruminant who consumes the same fruits will gain nothing directly. The sugar and starch will never make it past the microbes in its forestomach. The fatty acids the microbes leave behind after extracting energy from sugar and starch by fermentation in a low-oxygen environment are the only beneficial by-products that will continue the journey. A ruminant is thus one step removed on the food chain from all the plant products it consumes.

By now it may seem that evolution made a grave error in stationing microbial habitat at the fore rather than the aft of any mammalian herbivore. Forestomach processing is a bad move if woody forage is all an herbivorous lineage may expect, and it is a wasteful way to process the fruits the plant world offers. Yet forestomach processing was independently invented at least twice within the artiodactyl lineage, and again in the tree sloths, some kangaroos, some rodents, and several primates.[16] Because evolution took this path repeatedly, we should look for an adaptive explanation (or explanations). One idea is that ruminants benefit from being able to eat and run. They can snatch a quick meal in dangerous habitat and then chew it later in safety.[17] That argument sounds plausible for the ancestors of deer, but it doesn't work for lethargic tree sloths.

An adaptive explanation that applies to all forestomach animals is that they alone can manufacture every B vitamin and every essential amino acid from the blandest fare.[18] They do not, of course, do so directly. Microbes provide the service, and the hosts simply absorb the products through their small intestines. In contrast, hindgut processors require more diversity in fodder, as they must consume at least a smattering of plants that contain the amino acids and B vitamins in short supply. Microbes in the cecum and colon are manufacturing these goods, but the hindgut host has limited access to these products at that point—unless it is willing to eat its own shit (as rabbits and many other small hindgut herbivores regularly do for exactly this reason). A horse may appear to be eating only grass out in the pasture, but it is also choosing prairie flowers—some that are rather toxic—to balance its diet.

Hindgut herbivores can get by on woodier forage than can forestomach herbivores, but in nitrogen-poor habitat, the forestomach contingent has the hindguts beat. A cow recycles nitrogen far more effectively than does a horse because microbial fermentation is located in advance of the small intestine. Horses and cows require nitrogen for building and regenerating their own tissues (DNA and protein both contain nitrogen), but microbial symbionts need even more. A horse with access to only nitrogen-poor fodder is starving its microbes more than itself.[19] When the microbes begin to starve, so, too, will the horse.

A horse that consumes the legume locoweed (*Astragalus* spp.) in the American West is desperate for nitrogen. Experienced range horses are not stupid, but they have no choice. Better crazy than dead. Symbiotic microbes within the roots of many legumes, including locoweed, "fix" nitrogen from the atmosphere, and thus the plant can use it extravagantly in its own tissues. On poor range, available nitrogen will be concentrated in poisonous legumes. Indeed, the nitrogen profligacy of the legumes manifests in the nitrogen-rich alkaloids this group famously produces for the sole purpose of repelling herbivores. A plant wants its fruits to be eaten, but it does not want to donate its leaves. Locoweed thus intends to drive horses crazy; it intends to deter.

A cow on the same range, however, may be able to ignore locoweed; it may not be desperate for nitrogen. Even if it is, a cow can consume more of the plant without suffering neurological trauma. Again, the reason is that the microbes are up front. They mine the alkaloid for its nitrogen; the molecular remnant can then do no harm. Microbes in the colon of a horse would be happy to provide the same service, but they don't get the chance. The alkaloid passes through the lining of the small intestine into the horse's bloodstream.

Any foray into comparative physiology rapidly becomes overcomplex. As we have seen, there are good reasons for a lineage to select a forestomach fermentation system. And there are good reasons not to. There are good reasons to defer cellulose processing to the hindgut, and liabilities as well. Is it any surprise that nature has brought forth both lifeways in abundance?

Strictly from the fruit's perspective, hindgut processors may be the winners. Hindgut animals have the opportunity to grind or spit a seed on only one pass. They move bigger objects through their digestive tract. Acidosis is no issue for them when contemplating an acidic or sugary fruit, and every bit of sugar and simple starch they consume feeds their own tissues. For all these reasons, a fruit might find heaven in a hindgut—if there were only a hindgut to be found.

There were many big hindguts in North America until thirteen thousand years ago, and many large forestomachs as well. The forestomachs who vanished from this continent at the end of the Pleistocene include camel, shrub ox, and yak, along with taxa of bison, goat, muskox, and cervid. If ground sloths were forestomach processors, then they would constitute the largest of that extinct group. Many forestomachs, however, survived the catastrophe: bison and muskox, moose and caribou, elk and several species of deer, along with mountain sheep, mountain goat, and pronghorn. All these taxa are alive, if not all well, in North America. Forestomach fermenters took a hit, to be sure, but nothing like what happened to the hindguts at the end of the Pleistocene.

Gone are the American mastodon, Cuvier's gomphothere, and all three or four species of mammoth. The equids (horses) vanished entirely too. Perhaps the notoungulates (like toxodon) and the liptoterns of South America were hindgut fermenters; fossil evidence is silent on this issue. What we do know is that none—absolutely none—of the hindgut megafauna remain in North America. All are ghosts. Today the biggest hindgut herbivore north of Mexico is a rodent: the American beaver. Tropical regions of the Western Hemisphere are only slightly better off; tapirs are hindgut processors, and three smaller species escaped the end-Pleistocene extinction.

Extinction was catastrophic for megafauna in North and South America at the end of the Pleistocene epoch. From a fruit's perspective, the loss was even more onerous. Some of the megafauna remain, but the hindguts have all but vanished.

Elephants as Proxy Mastodons

"Affie may be the first elephant to ingest fruit of *Maclura* in over 10,000 years!" Paul Martin exulted. He emailed me the wonderful news shortly after the turn of the millennium. With his connections and my ability to scavenge fruit, we had succeeded in putting one of America's extreme anachronisms within reach of an elephant proboscis.

Actually, three elephants were given a chance at an osage orange. These were the African elephants at Chicago's Brookfield Zoo. Zoos aren't in the business of experimenting on their animals, as I learned when I volunteered to deliver a gift of avocado and pawpaw to the elephants at the zoo in Albuquerque, New Mexico. Those elephants were accustomed to watermelon, cantaloupe, apples, oranges, and (right after Halloween) pumpkins. Veterinary staff would not have their animals exposed to unfamiliar fruits without a very good reason. And my reason—to satisfy my own curiosity about putative anachronisms—was not good enough.

Paul Martin has connections that I lack: he knows the director of the Brookfield Zoo. As well, Paul's enthusiasm is of no minor assistance. I had approached the Albuquerque Zoo with the utmost sobriety and professionalism. Paul is famous enough to get away with communicating feelings as well. To George Rabb, the director of the Brookfield Zoo, he wrote that if the elephants were to consume maclura, "Such a discovery would shake up botanists who have not learned the delights of time traveling into the wild world of the late Quaternary."

Rabb had his staff search the literature for a better understanding of maclura, and I provided information by email. A determination was made that the toxins in maclura were not of the sort that would injure an elephant, particularly in the dose of a fruit or two. The experiment went ahead. Here is what happened:

Affie, the matriarch of the Brookfield elephants, did eat maclura—but just the first fruit she was offered. After that, she showed no interest in any more. The reactions of the other elephants were strongly negative. One wasn't even willing to smell the fruit when the offer was first made. Finally, she took it from her keeper and hurled it down the hall. The second elephant did the same thing but aimed for the public area.[20]

How to interpret these results? At the outset, Paul and I had reckoned that there was nothing the zoo elephants could do that would falsify our hypothesis that maclura was an anachronistic fruit evolved, at least in part, to attract the dispersal services of proboscideans. Herbivores are known to be wary of novel foods. Cultural knowledge of gastronomic possibilities, passed from one generation to the next in social animals, would not be available to zoo elephants. Then, too, we didn't know enough about maclura to judge when it was ripe. Does it need frost to ripen, like American persimmon? Most important, we knew that well-fed animals are no substitute for wild creatures. Many fruits, including maclura, become available to animals in a season of want—a dry spell in the tropics, autumn and winter in the temperate zone. Anything that can help an animal fatten up for the months of deprivation may be very attractive, especially if it occurs in bulk beneath the canopy of a tree that one can remember to visit year to year.

This experiment was not, therefore, a rigorous test. Paul and I were just immensely curious. Ideally we would have lobbied for the experiment to happen over a number of days, with the elephants exposed to the fruits more than once. Affie did, after all, exactly what a savvy wild elephant would have done on first contact with a strange-smelling fruit. The right response is to try just one. If there is no digestive upset or other physiological problem, then in a few days perhaps more can be eaten.[21]

It hadn't occurred to me then, but now I think that the other elephants did exactly the right thing too. Let the matriarch do the sampling, and if you see her take additional fruit on another day, well maybe you can eat them too. Follow your leader.

Paul and I were thrilled by the results. Dan Janzen wasn't. "Irrelevant," he told me in an email. "Zoo tests mean almost nothing. The wild animals that we have long been around [in Costa Rica] have grossly different behaviors than their congeners in zoos. Same for humans in jail."

Testing a candidate anachronism on a zoo elephant may be irrelevant, but it is nevertheless irresistible. Janzen had performed such a test himself six months after his initial insight that some fruits in the Americas might have lost their dispersal partners. Janzen invited me to rummage through his anachronism files at the University of Pennsylvania while he was away in Costa Rica. The files contained a great deal of correspondence, reprints, clippings, and this: "First trial feeding of elephants in the Philadelphia zoo." The one-page document was dated 30 March 1978. The report indicated that three kinds of Costa Rican fruits were offered to the zoo's four elephants.

The lone Asian (Indian) elephant ate as many *Guazuma ulmifolia* fruits as her keeper made available. Two of the African elephants ate several and then showed no further interest. The third African elephant added a couple of fruits to a mouthful of hay she was chewing, then spit out the whole mash and would not sample any more. The keeper had never before seen this elephant spit anything out.

All four elephants crushed the tough-husked spheres of *Crescentia alata* with their feet, but all were then repelled by the odor—which Janzen reports is intensely fetid by the time range horses judge this fruit acceptably ripe. The next fruit tested was the legume pod *Cassia grandis*. The Indian elephant took the pod she was offered and broke it neatly in half, holding one end in her trunk while smashing the other against a foot. She discarded part

DUNG OF A GROUND SLOTH. This dung bolus of the extinct Shasta ground sloth, *Nothrotheriops shastensis,* was recovered from a dry cave in Arizona. Paul Martin for scale.

and squashed the rest underfoot, ignoring the residue from then on. Two of the African elephants wanted nothing to do with the pods they were offered. The third deliberately squashed one and then disregarded it.

Another group of elephants treated an unfamiliar legume pod in much the same manner twenty-one years later. These elephants were not zoo animals. They had spent their lives performing in circuses but were now comfortably retired on an estate in Tennessee that had been developed in 1995 expressly for their well-being. The Elephant Sanctuary consists of a heated barn plus 112 acres of fenced grounds that the elephants wander at will, encircled by a much larger wooded area.[22] Paul Martin had made contact with the director, Carol Buckley. I followed up, sending her information on wild fruits eaten by elephants in Africa, along with reports on the widespread use of honey locust pods as livestock fodder. She agreed to offer honey locust to "the girls."

I sent a box of pods harvested from my ranch neighbor's young trees. Several weeks later, I received this report: "None of our girls ate the pods, but they all handled them. They picked up the pods immediately and rolled them in the palm of their trunk until they were pulverized. [Honey locust breaks apart easily, unlike the thick woody pod of *Cassia grandis*.] They then dropped or threw the pulverized pods to the side. They showed no further interest in the pods after that. I offered pods a second time and they would not even smell or touch them. It appeared to me that they crushed the pods initially to see what was inside." The negative result was compounded by more information Carol Buckley provided. She told me that there are wild persimmon trees on the grounds, but the elephants step on the fallen fruit—even after a frost—rather than eat them. Worse, they spend a lot of time delicately plucking and eating a fruit that is not supposed to be megafaunal fare at all: blackberries.

My vicarious experience with the retired elephants was not a total bust, however. The sanctuary elephants had developed—rather, resurrected—a curious behavior that would have profound implications.

Let Them Eat Clay

"One last question," I wrote Carol Buckley in an email. "Is there any evidence of the girls eating dirt, now that they have a big area to roam? Any places where the dirt is muddier like clay, rather than sandy?"

"Yes, the girls do eat dirt," she replied. "Clay is what they prefer. In the winter they eat more than at other times of the year. They don't eat handfuls but pick around for a choice morsel."

For two reasons, this response filled me with excitement. First, I had begun to suspect that, absent a clay source, the matriarch of the Chicago elephant herd would be wise to refrain from making maclura a regular part of her diet.[23] Second, the behavior of the Tennessee elephants suggested that no cultural knowledge may be needed for animals to resurrect an aptitude for detoxifying food. The behavior comes naturally, or perhaps it is remembered. All the retirees at the Elephant Sanctuary had been born in the wild. They spent anywhere from six months to three years with their birth families before they were captured and trained to participate in circus acts. Some of these elephants thus may have watched their elders select and consume clay. Some may have been old enough to experiment with clay themselves. Perhaps some even learned which foliage and fruit should be taken with clay and which could be eaten without. They may have learned to taste which soils were rich in clay. Subsequent decades spent in concrete habitats should have dulled the memory. But then, elephants are renowned for their recall.

Only recently have scientists begun to realize how important clay may be for wild herbivores. Herbivores throughout the world have long been known to visit well-trampled spots or caves to eat dirt, but the assumption was made early on that the attraction was mineral concentrations such as sodium, calcium, and elements that animals need in trace amounts. Places that herbivores frequented to eat dirt were thus called mineral licks or salt licks. The name has misled researchers ever since.

Plant eaters often are attracted to minerals. Ranchers provide salt blocks for their livestock; rodents gnaw on bones. But minerals may not be the only supplements animals require. Within the past five years, field ecologists have increasingly reported that many of the so-called mineral licks contain concentrations of vital minerals no greater (and sometimes less) than surrounding, uneaten soils. Indeed, the only universal feature of mineral licks throughout the world is a very high clay content.[24]

Why do hindgut herbivores like elephants, rhinos, zebra, and horses travel considerable distances to consume clay? Why do ruminants like giraffes, kudu, eland, antelope, water buffalo, and duiker engage in geophagy too? Why do chimpanzees, gorillas, various monkeys, and baboons of both sexes ingest clay, sometimes daily?[25] Why do parrots make long daily flights to clay licks? Why do tapir, grouse, rabbits, squirrels, koala, tortoises, and many other small herbivores consume clay-rich soils locally available?[26]

Clays are weathering products of alumino-silicate rocks—meaning everything except limestone. Negatively charged, their platelike structure gives them high surface-to-volume ratios ideal for adsorbing many kinds of toxins, ranging from alkaloids and oxalates to the powerful toxins emitted by molds and bacteria that rot food. (Molds and bacteria produce rot toxins for the express purpose of persuading vertebrates like us to leave the food alone once the microbes have staked a claim.)[27] Clay deactivates toxins not by dismembering them but by grasping them in such a way that digestive processes cannot liberate the ill effects. The toxins are then defecated along with the clay.

A casual vegetarian, I sometimes forget that plants are not on this Earth to be my friends. They did not evolve to rest benignly on a dinner plate. For hundreds of millions of years, leafy plants have engaged in arms races against the big and little herbivores who regard them as lunch. Thorns are the visible products of this extended warfare. Alkaloids, tannins, saponins, resins, hemagglutinins, proteinase inhibitors, and impostor amino acids may not be visible, but they are intended to be sensed. Many plant poisons, thankfully, taste terrible. A plant does not, after all, want to slip something past us. It wants us to exclaim "ugh!" after just one bite. Intense bitter alerts us to stay away from quinine and many other alkaloids, including the caffeine and theobromine in unprocessed chocolate. I love the fragrance of some cyanide-producing compounds, which provide the almond scent not only of almonds but of seeds within the fruits of other members of genus *Prunus*, such as plum and apricot (medicinal laetrile is extracted from the latter). Cyanide toxins are also abundant in the seeds of apple, which is a related genus *(Malus)*. Were I to overdose on such plant products (a tablespoon or two of ground apple seeds would probably do it) my reptilian brain would promptly switch the verdict. What had once been fragrant would now repel.

Tasting something the second time around (a euphemism for vomiting) is enough for humans to develop food aversions to almost anything. Dan Janzen tested more than three hundred species of leafy plants on a captive tapir (a hindgut processor). For some species, the tapir would refuse even to sample a leaf; one whiff was sufficient to detect a toxin at intolerable levels. But if the tapir ingested and then regurgitated a noxious plant, he would never eat it again. Janzen discovered that even for foods the tapir seemed eager to consume the animal would stop after "a few percentage of the tapir's stomach capacity" and refuse to eat again until offered something else.[28] Tapir thus lives by the dictum "Everything in moderation." Poison is, after all, in the dose. And some poisons (tannins and alkaloids, for example) may cancel one another out.

Some of our most important medicines are manufactured by plants for the express purpose of deterring herbivores like us. In small doses, however, they provide humans (and, we are just beginning to learn, other primates who use plants medicinally) with gifts from the gods. Aspirin is derived from the bark of willow, and quinine-based antimalarial pills are prepared from the bark of the South American cinchona tree. From kin of garden tomatoes, we obtain atropine and scopolamine—both of which were part of my mother's homecare hospice regimen, along with the most blessed of all bequests from the plant realm: morphine, a derivative of the seeds of opium poppies. Opium, cocaine, nicotine, caffeine, theobromine—all are alkaloids manufactured by plants to deter, not addict. "The caffeine in the cup of coffee that gets you going in the morning," writes Susan Allport, "is first and foremost an effective insecticide."[29]

Plants equip their greenery and especially their nutrient-rich seeds with toxins. Seeds may not demand much if any chemical protection if encased in a nearly impenetrable shell, as are walnuts. Soft seeds, however, may be arsenals of the deadliest poisons of all: strychnine, cyanide, opium, mescaline, peyote. Beware the avocado seed. Beware, too, the weakly shelled cashew and the almond—at least in the raw state. Because roasting can deactivate many toxins, humans have raised the stakes in this particular arms race. It is only because humans possess the gift of fire that we can make these and many other seeds not only edible but edible in bulk. For example, dried beans and lentils would be lethal if eaten raw by the very people who now rely on these seeds as staples. Roasting will not deactivate bean toxins, however; only moist heat will work.[30] When foods are transferred across cultures, culinary wisdom must be part of the exchange. Exotic foods require exotic cuisines. I cringe when walking past bins of raw cashews, raw almonds, and (gasp!) raw apricot kernels at health food stores. How many unsuspecting customers are straining their systems or even killing cells by snacking on these unprocessed (and thus presumably more healthful) foods? For how many meals did I myself prepare lentils improperly?

Thousands of years of intentional breeding have diminished toxic components of domesticated plants. The kinds of potatoes I eat contain just a fifth of the concentrations of glycoalkaloids that Andean peoples encountered in the ancestral stock.[31] One-fifth of a toxin is, however, still a toxin. That's what our livers are for. Blood circulated through the absorptive linings of the stomach and intestines passes through the liver before going anywhere else. Our livers evolved to dismantle poisons. We needn't shield them from plant aggression entirely. Nevertheless, we should take care not to overburden them. (That's why nonfatal poisonings so often result in irreversible liver damage.) Everything in moderation. You can safely eat almonds that have been roasted—but, even then, not too many at a time. The same holds for boiled legumes, potatoes, spinach—just about anything we eat that is not meat, dairy, cereal grain, or fruit pulp. From an evolutionary perspective, we know that fruit pulp, nectar, and sometimes pollen are the only parts of the plant intended to be eaten. The plant realm offers fruit pulp to the animal realm with alacrity. The pulp should be toxin-free, shouldn't it?

Not always. Maclura is an instructive exception. Why is the pulp of maclura too toxic for humans or their pampered livestock to consume? Why are zoo elephants leery of it?

Plants have two compelling reasons to poison their fruit pulps. First, unripe fruit is never supposed to be eaten. It does a fruit no good to be consumed before the seeds are ready for a life on their own. Even our best-loved fruits are unpalatable if taken too soon. A green banana tastes like chalk. Underripe apples and persimmons pucker the mouth painfully. Chalkiness and puckering signal the presence of astringent compounds, usually tannins, which are a favored way for plants to keep their developing fruits safe. The plant can easily deactivate the tannins, when the time is right, by adding proteins or particular kinds of carbohydrates to the mix. Tannins naturally bind to these substances (hence their use for "tanning" animal skins, binding the molecules of the hide in such a way that microbes will not be able to extract food value from the leather). The new compound molecules can then pass harmlessly through a gut.

A second reason for plants to poison their fruit pulp is to repel insects. Ants and dung beetles are the only insects known to be helpful in dispersing plants. The rest of the arthropods are enemies. Seeds are often at or near the center of a fruit, distant from insects who might lay eggs on the outer skin or rind. Hatched larvae must tunnel through pulp to get to the protein-rich seed. If that pulp is acidic (lemon), capable of tenderizing meat (papaya), or imbued with sticky latex (maclura), the larvae die en route. Many of the things we vertebrates are equipped to eat would be irritating or damaging if we had to swim and breathe in them.

On the whole, however, insects and vertebrates are vulnerable to many of the same poisons. Proteinase inhibitors will arrest the deployment of enzymes necessary to digest proteins, whether those enzymes are produced by caterpillar or camel. Impostor amino acids will volunteer to be woven into a protein, whether that protein is within the cell of beetle or buffalo. The one thing a vertebrate can do that an insect cannot (at least while the insect is still a flightless larva) is visit a clay source. Humans can do even better by cooking plants in clay pots.[32] Once vertebrates discovered the benefits of clay, plants had an opportunity to escalate the arms race against insects without turning away their valued mutualists. Maclura could stock its fruit pulp with toxins, so long as the toxins were of a kind that could be deactivated by clay.

That evolutionary scenario is a hypothesis, of course, and one that I cannot corroborate in the literature.[33] The rethinking of mineral licks as clay licks is too new for scientists to have fully developed the implications. What does seem certain is that many tests of the palatability and toxicity of wild plants that have been performed on captive animals will have to be redone. If an elephant or tapir cannot access a clay source, we cannot ascertain its food preferences in the wild.[34]

Another implication yet to be explored is what the clay connection may mean for zoos. In 1995 scientists not only confirmed geophagy in mountain gorillas in Rwanda but linked the practice to the season in which gorillas depend on bamboo shoots as their primary food.[35] Bamboo shoots are known to contain high levels of cyanogenic (cyanide-producing) compounds. Clay may adsorb those toxins. This finding should be of special interest not only to zoos that house gorillas but also to those caring for other bamboo eaters: the giant panda and the red panda of China. The Central Park outpost of the Bronx Zoo has had some success in breeding the endangered red panda. I have watched those pandas plod around the outdoor grounds, and now I wonder whether the panda habitat is fortuitously positioned on clay-rich soils. Perhaps this is why captive breeding programs for a variety of endangered mammals have had more success when the animals have been given a habitat in which dirt rather than concrete is underfoot.

If clay is a beneficial dietary supplement for herbivorous vertebrates, it may become even more beneficial when an animal is pregnant. I have found no reports of a link between geophagy and pregnancy in the wildlife literature, but one very recent anthropological study offers tantalizing clues. A 1998 paper by Andrea S. Wiley and Solomon H. Katz, titled "Geophagy in Pregnancy: A Test of a Hypothesis," reports that in rural Africa it is commonly believed that one of the first signs of pregnancy is a craving for clay. If a pregnant woman cannot get out to the termite mound for the best clay, she will snack on the mud in the walls of her hut. (Termites consume clay they mine from tens of meters beneath the ground, possibly to detoxify the lignin in their woody diet as well as to turn their feces into adobe pellets for building the mounds.)[36] Clay is believed by the African women to reduce morning sickness. What is morning sickness, with its attendant food aversions, if not a means to ensure that the fetus is spared exposure to plant hormones and toxins when most vulnerable?[37]

Geophagy is common to many cultures.[38] In America, however, the tradition continues only among women of African descent in some parts of the rural South.[39] If geophagy is healthful for the fetus and palliative for the mother, why has the practice been abandoned by urban women in my country? Perhaps one reason is that even the poorest among us can switch to a diet rich in meat, dairy, and relatively nontoxic plants whenever the urge arises. Women in poorer countries may not have this luxury. Perhaps, too, because women (and sometimes children) are the only clay eaters in some cultures, the men who charted the course for Western medicine and psychology chose to treat the practice as an aberration. Perhaps geophagy fell out of favor when the opinion leaders of my culture began looking through microscopes, finding that soil teems with little worms and other vermin, becoming obsessed with cleanliness. New York City, after all, would be impossible without modern sanitation and awareness of proper hygiene. But carefully selected clay is not mere dirt, and a pill of supplementary minerals is no substitute. Nevertheless, psychologists have long regarded geophagy as "aberrant, desperate, or unseemly behavior."[40]

A closer look at geophagy as practiced by humans shows that it need not involve ingestion of anything that would do us harm. The African American women who collect clay extract it at depth. In Africa, termite mounds may be relatively free of parasites because the mounds are elevated above the ground that is trampled and defecated on by animals attracted to the clay. Tradition sometimes calls for clay to be baked before consumption.[41] (One wonders whether this practice could account for the serendipitous discovery of pottery making.)

On a lark, I asked the women in my family who are mothers whether they had craved clay during their pregnancies. Two said no, but the third told an astonishing story. It begins routinely enough: "I just went berserk with salty stuff. Topor's kosher dill pickles, sauerkraut, and ham topped with cheese and mustard were my faves. I also seemed to like to drink lemonade with all this salty stuff." Where my sister's story gets interesting is in the addendum: "I do not know if you recall this or not, but I specifically remember as a child craving Kaopectate. I absolutely loved the taste of it, often feigning illness to get my fix."

Whoa! Kaopectate used to contain kaolin, a kind of clay popular among clay eaters, and it now contains even more adsorbent clays. We routinely use it as medicine when recovering from a mild dose of food poisoning. We do not, however, regard clay in any form as a dietary supplement for women in the early stages of pregnancy, nor for healthy children. Almost all of the zoological reports I have collected on animal geophagy that are less than five years old routinely liken the chemical assay of the clay lick in question to the composition of Kaopectate.

I haven't made Kaopectate a regular part of my diet, and you shouldn't either. The clay now in Kaopectate is not one of the types usually associated with human geophagy. Also, too much clay ingested and with the wrong foods may deprive the body of dietary calcium. Perhaps the day will come, however, when a shaker of powdered clay appears on the table along with the salt. Until then, heed the wisdom of children. Children who refuse to eat a new vegetable are, despite what their parents think, doing exactly the right thing.[42] Parents might nevertheless encourage their young to follow the lead of Affie the elephant: take just one bite and see how you feel.

Or maybe not. Acquaintance with the toxin-herbivore literature makes me suspect that children's widespread aversion to green vegetables like asparagus and spinach is downright healthy. Their youthful livers may not have developed sufficiently to process such foods, and the oxalates in spinach may rob them of calcium at the time they need it most.[43] Spicy foods? Not child's fare either. Most spices are extracted from seeds or bark—and they were put there expressly to dissuade herbivores from treating such vital plant parts as food.

In rural Africa young children will stop by the termite mound for a hit of clay on their way to school. I bet I did too—though not precisely the termite mound. I remember being very fond of making mud pies, the real kind, and I would not be surprised if I "inadvertently" put a finger in my mouth while doing so. Perhaps the story my mother used to tell of catching me eating a worm needs revision: I was probably just eating dirt, and the worm happened to get in the way.

Eating clay is such a natural activity for an omnivorous primate that, until kids were stuck in concrete habitats, our species probably got by just fine despite cultural loss of the tradition. Ideally our mothers would have poured a mud sauce onto our food (as Germans used to do), kneaded clay into bread dough (a Swedish tradition), mixed clay with acorn mash (a culinary custom of many Native American tribes and the Sardinians), and prepared a fondue of clay for dipping our boiled potatoes (still done in the Andes).[44] Then again, if my mud pies were sufficient, why did my sister and I eat wallpaper when we were toddlers? Perhaps I was still so young that my mother was obsessively watchful for every fistful of dirt that might approach my mouth. Or perhaps I practiced plasterphagy only during the winter, when I had no access to soil. Why, too, do ghetto kids ingest lead-contaminated paint chips? Might these habits have something to do with the fact that paint and wallpaper glues contain clay to prevent dripping? Why did my sister crave Kaopectate—and did that craving cease during the summer in our ungroomed suburban environment, with its fields and ditches and dirt roads? I remember a lovely blue clay layer that would appear if we dug too deep in the garden. Why do I remember it so well?

Speculation can grow wild: Might childhood allergies have something to do with the fact that kids don't play in the mud anymore? Are nursery schools that provide clay for playtime helping more than the artistic growth of children? Do urban elders afflicted with osteoporosis have something to learn from women of the rural South?

Enough. For now we need consider nothing more than that the extinct megafauna of North America who evolved eating maclura, gymnocladus, desert gourd, and other anachronistic fruits with poisonous pulp had long supplemented their inborn detox systems with a gift of earth itself.

THE ECOLOGICALLY WIDOWED. The yellow fruits of *Solanum elaegnifolium* contain toxic alkaloids that deter dispersal partnerships with mammals but not with reptiles. Wherever box turtles are locally missing, these fruits are overbuilt for dispersal.

9

Consequences

WHAT HAPPENED AT THE END OF THE PLEISTOCENE was not a mass extinction in the usual sense. Although hippolike notoungulates vanished from South America, giant marsupials from Australia, and proboscideans from most of their range, smaller mammals carried on as before. While the biggest vultures and other carrion-feeding birds spiraled into extinction, songbirds were unaffected. Nary a loss ripped through Earth's most extensive life zone, as sharks and whales continued to patrol the seas and fishes and foraminifera enriched the shoals.

Not so today. The little ones are losing ground, along with the remaining megafauna, and the seas are everywhere under assault. Paul Martin believes that the extinctions of the last forty millennia represent "the lethal overture to the sixth great mass extinction, currently under way."[1] The slate of modern-day causes is depressingly familiar: overhunting, habitat loss, invasion of exotic species, fragmentation of landscapes, pollution. At the root of each cause is us. In this century and beyond, another cause will prominently join the list. Dan Janzen calls it "ecological ripples."

"All species that I have worked with in detail in the tropics have the property that the demise of one of them will cause an ecological ripple in the habitat," Janzen writes. "Some species make bigger ripples than do others when removed, but even this trait will vary in the eye of the beholder."[2]

Ecological interactions are as endangered as the species themselves. Stephen L. Buchmann and Gary Paul Nabhan recently sounded the alarm for "the forgotten pollinators"—the insects, birds, and bats who depend on the nectar or pollen of flowering plants for food, and the plants who, in turn, depend on these animal mutualists for help in conceiving the next generation.[3] Nearly as important is *launching* the next generation.

What Paul Martin calls "extinction of the massive" at the end of the Pleistocene put at risk plants who depended on megafauna to disperse their seeds.[4] The fruits of these plants are burdened with traits that are now anachronistic. Much is to be learned from this episode of ecological ripples. What have been the consequences?

Ecological Ripples and Extinction Cascades

The record of fossil plants is a fragment of the botanical tapestry woven in space and time. Preservation of plant materials, especially in upland or mountainous habitats, is a rare occurrence. The only record of a plant extinction in the very late Pleistocene is that of a spruce tree in North America.[5] This does not, however, imply that only a single tree was lost. New taxa of living plants are discovered virtually every time a botanist visits an uninventoried tract of tropical forest. New species are documented even in the United States, and a wholly new genus of shrub growing in the mountains of southeastern Arizona was described just thirty years ago.[6] And yet the discovery of any new plant is almost always announced as a new taxon—not as a fossil plant rediscovered.[7] The paleobotanical record is far from complete.

The avocado and papaya, pawpaw and cherimoya, osage orange and desert gourd, honey locust and Kentucky coffee tree survived the loss of their target dispersers at the end of the Pleistocene. Perhaps other large-fruited plants vanished without a trace. This is what Dan Janzen suspects. He and Paul Martin disagreed on this issue when they wrote their 1982 paper.[8] The tropics, where Janzen works, is a notoriously poor environment for long-term preservation of plant parts. Pollen is by far the most common botanical fossil, but it is prolific only for wind-pollinated plants, such as spruce. Even sources rich in plant macrofossils—stems, leaves, seeds—have not been carefully searched for trees suspected of losing their natural dispersers.

Martin has a long acquaintance with the record of fossil pollen preserved in bog sediments of eastern North America and trapped in the middens of packrats of the desert West. It was during the late 1950s, while he was spending his days sorting and counting fossil pollen, that Martin began to develop his megafaunal extinction and overkill ideas. Today he cautions, "Paleobotanists do recognize severe extirpation and extinction of trees in the Pliocene–early Pleistocene of Europe and somewhat earlier in North America. But to my knowledge very few discoveries of extirpations or extinctions are known in the last hundred thousand years in North America."[9]

Although little is to be found in the fossil record, insects may have been lost as well when the megafauna disappeared. Some bruchid beetles alive today depend on the seeds of a single species of tropical fruit for the growth of their young.[10] If that fruit goes, presumably so goes the beetle. Insect species lost in an ecological ripple ten thousand years ago may remain unknown forever. Only a half dozen species of Pleistocene dung beetles were fortuitously preserved in the La Brea tar pits of California, along with the bones of ground sloths, sabertooth cats, dire wolves, and carrion-feeding birds. Yet of these six beetles, two (possibly four) are extinct.[11] The dung beetle extinctions are not surprising; when the megafauna vanished, beetles who nurtured their young in dung were even more vulnerable to extinction than were the large-seeded plants who depended on animals for transportation.

Dan Janzen studied beetles that scavenge horse and cow dung in Costa Rica. He concluded, "The Santa Rosa dung beetle fauna is probably a mere remnant of what was once supported by the Pleistocene megafauna. With the extinction of ground sloths, gomphotheres, glyptodonts, horses, etc., the dung beetles that survived would have been those that could survive on the very diffuse and small-particle dung rain that is generated by humans and the many small species of vertebrates in a tropical deciduous forest. With the introduction of a cow and horse megafauna, the dung beetles have again a resource that comes in one- or two-kilogram packets in large numbers. However, it should be a long time before the species richness again climbs to the level that was probably supported by the Pleistocene megafauna, since there is no nearby source area with a rich fauna of large dung beetles from which more species may immigrate."[12]

Australians could not wait for evolution to work its magic. Owing to a booming cattle business, populations of biting flies exploded, and mummified cow pies crowded out range grasses. Drastic action was called for, so dung beetles were imported from Africa and the Mediterranean. Of the forty-one species of beetle introduced beginning in 1968, twenty-two took hold and many of those thrived. The benefits have been substantial; drawbacks, if any, have not been reported.[13]

The United States has a dung problem too, especially in heavily grazed pastures of the arid West. A dozen or so species of dung beetle have been introduced from the Old World (most, unintentionally), but these are the smaller forms, not the great scarabs of Africa.[14] Even in Africa, however, dung beetles cannot process feces dropped during the dry season. Only termites can do that job. One wonders whether the Americas may have lost a vibrant termite ecology in arid lands at the end of the Pleistocene—and whether a careful study of restoration prospects should include termites. Perhaps our termites went extinct when animals, who had long been attracted to the mounds as sources of clay, became ghosts and thus could no longer deliver food-rich pellets and pats to the doorsteps of their insect partners.

Another ecological ripple that got its start thirteen thousand years ago may explain, in part, retrenchment in the range of North America's only surviving condor to coastal habitats along the Pacific. There the condor continued to find megafaunal carrion, in beached carcasses of whales and sea elephants.[15] In recent times, lead shot ingested from scavenged meat of lesser creatures seems to have been the cause of the condor's demise, but the unsuitability of wild carrion may continue to thwart reintroduction programs.[16] Perhaps the condor relied on more formidable scavenging birds (the great teratorns, extinct at the end of the Pleistocene) to penetrate stiff hides—just as slender-billed carrion-feeding vultures in Africa today depend on the stout bills of the lappet-faced vulture and the white-headed vulture.[17]

"To behold the Grand Canyon without thoughts of its ancient condors, sloths, and goats," Paul Martin has written, "is to be half blind."[18]

Even if we limit our sights to the precarious future of Earth's remaining seed dispersers, the consequences are frightening. An ecological ripple may become an extinction cascade.

Dan Janzen cautions that if the guanacaste tree were to vanish, nectar-feeding bats might too. Guanacaste trees in Costa Rica bloom during more than a third of the year. In a concatenation of effects, extirpation or extinction of bats that depend on guanacaste nectar would threaten survival of other flowering trees dependent on bats for pollination.[19] Some of those trees might produce fruits that frugivorous monkeys and birds rely on when other fruits are not in season, and so on. Henry Howe warns that local extinction of the *Casearia* tree at his study site in Costa Rica would almost certainly put several birds (toucans and cotingas) at risk in "a widening circle of extinctions."[20]

Neotropical forests are now challenged by a second wave of disperser extinctions. Tapirs are the only hindgut megafauna in wild neotropical forests today, and they are crucial for dispersing big-seeded trees.[21] Alas, they are easily hunted into "ecological extinction." The job then falls to the large birds and primates (woolly and spider monkeys are especially important mutualists). But these, too, are vulnerable to near extirpation by burgeoning populations of hungry peasants—even as the trees are secured from logging in designated reserves. The resultant creeping loss of tree diversity within what Kent Redford calls "the empty forest" will bear consequences for generations to come.[22]

Fruiting plants dependent on birds are found throughout the world. Two trees in New Guinea are dispersed only by birds of paradise.[23] The cassowary is the sole disperser of many large-seeded fruits in Australia.[24] Such plants may be as tied to their mutualists as the tambalacoque tree was to the dodo. In Africa, hornbills rank among the most important seed dispersers of the tropical forests, and "they are likely to become increasingly important in forest regeneration as populations of larger mammalian seed dispersers (such as forest elephants and primates) diminish."[25]

Gorillas and chimpanzees are vital contributors to plant diversity. These apes swallow much larger seeds than will monkeys. Gorillas and chimpanzees eat many of the same fruits, but there is one tree in the chocolate family (Sterculiaceae) that only a gorilla can disperse. This is *Cola lizae,* a plant recognized and named by scientists only in the 1980s.[26] Its seeds are too big for a chimpanzee to swallow. *Cola* fruits of all species—including the kinds of seeds that flavor cola drinks—are unusual in that elephants avoid them, but they are preferred foods of gorillas who know how to handle them. To eat *Cola lizae,* a gorilla must first pour off the poisonous liquid—a task an elephant presumably cannot perform. Then the animal must consume the pulp and swallow the entrained seeds without damaging the cotyledons exposed on the seeds. The crimson color of the cotyledons is a warning, as each contains a punishing dose of caffeine.[27]

In Chapter 3 we learned of a number of African trees dependent on elephant dispersal as well as an Asian tree dependent on rhinos. Asian forest ecology is poorly understood, but enough is known to sound the alarm. In a wide-ranging review of vertebrate seed dispersal in tropical regions of Asia, Richard Corlett judged that "loss of dispersal agents may, in the long term, be as serious a threat to tropical plant diversity as deforestation."[28] In a sad reminder of the rotting fruit that caught Dan Janzen's attention, Corlett writes that in tropical Asia, "Piles of uneaten fruits, rotting under the parent tree, are a characteristic and depressing feature of forests which have lost all their large vertebrates."

Regeneration of large-fruited trees in Madagascar may already be hampered by the recent extinction of a gorilla-size lemur. The largest remaining lemur, *Eulemur fulvus,* is now crucial for the continuance of a fully fruited forest.[29] In southern Africa there is no doubt that a wild cucumber utterly relies on a single animal, the aardvark, for dispersal. Zoologists have wondered why, alone among animals preying on ants and termites, aardvarks retain functional cheek teeth. *Cucumis humifructus* may be the reason. This strange vine produces globose fruits five to nine centimeters in diameter, which grow on stalks that rise and then arc downward into the soil for thirty or more centimeters. Aardvarks excavate the ripe fruits for water as well as food. Seeds are not only defecated intact but also planted by aardvarks, as the animals bury their feces.[30]

THE GIANT FLOWER OF RAFFLESIA. One of the strangest of all tropical plants, *Rafflesia* is endangered by the loss of its Malaysian pollinators as well as its seed dispersers.

Reptiles may be important fruit distributors, especially on islands lacking large mammals. A number of New Zealand plants produce fruits of a design that might well be regarded as a lizard dispersal syndrome.[31] Such fruits are sweet, less than 5 mm in diameter, and striking for their color: they are white, translucent, or pale blue or pink. Equally telling is their placement on the plant. Many of the divaricating shrubs discussed in Chapter 7 produce little fruits within dense tangles of branches that flighted birds cannot penetrate. *Hymenanthera alpina* is an extreme example. The white berries of this prostrate plant are borne on the underside of stems that hug the ground.

In addition to the moa, New Zealand lost its largest lizards during the past thousand years of human occupation. Bones of an extinct giant gecko, discovered in the 1980s, confirm the presence of a reptile large enough to have consumed the karaka fruit, whose seed is too big to be swallowed by any living animal native to the islands. Very large geckos that are not yet extinct on the island of New Caledonia to the north are highly frugivorous. Karaka fruit is thus an anachronism and a giant gecko its ghost. Other New Zealand plants dependent on lizards for dispersal may be anachronisms in the making. Rats and domestic cats accompanied human colonists and have been plundering lizard populations ever since.[32]

Of all plants that beg for explanations of adaptation and rarity, the fourteen species of genus *Rafflesia* (family Rafflesiaceae) are the strangest. All are parasitic on—rather, within—vines of Southeast Asian forests. *Rafflesia* has no visible vegetative parts. Only the flowers emerge from the stems or shallow roots of their hosts. And what flowers they are! The color and odor of carrion, these flowers are colossal.

Each *Rafflesia* species is known from only a few sites. This is obviously a lineage in distress. The genus bears the stigma of anachronism in two debilitating forms: in its mode of pollination and dispersal. Male and female flowers are borne on separate plants, so the species absolutely depends on pollinators. Yet flies attracted to the odor are ineffective in transferring pollen. One wonders if a larger carrion or dung beetle, now locally extinct, might have been the evolutionary target. And one wonders if that postulated beetle disappeared for want of elephantine carrion or dung.

As to dispersal, small rodents that consume bits of pulp, swallowing the tiny seeds whole, are the only effective agents available to *Rafflesia* today. For *Rafflesia keithii,* Louise Emmons and her colleagues judged such rodents to be "ideally suited" as seed dispersers.[33] She does mention previous hypotheses that Asian elephants might once have been the match, and she writes approvingly of the Janzen and Martin anachronism paper. Nevertheless, squirrels are her choice. Part of Emmons's reason for siding with squirrels is the concern that elephants and other megafauna would deposit too many of the tiny seeds in a single dropping. Of the hundreds that would likely be deposited in one bolus, only a few or even one seed could grow to maturity at the site. This argument is sound as far as it goes, but it does not go far enough. We must consider the missing dung beetles. In forests where elephants still thrive, fresh dung may be dismembered within a few hours.[34] Some bits are drawn down into burrows excavated directly beneath the source, but other scarabs roll grape- or plum-size treasures for many meters. Consider the authors' description of this fruit; everything about it, save perhaps the size of the seeds, is compellingly megafaunal:

> The collected fruit of *R. keithii* was a sessile ball about 14 cm in diameter, a giant berry. Its outer surface was dull, blackish-brown and patterned with large, roughly pentagonal sections separated by grooves. . . . Before it was opened to show the pale interior, it was extremely cryptic among the leaf litter on the ground. The outer covering was barklike, woody, and quite hard. The ball was filled with smooth, oily, cream-colored flesh with a viscous

> custard-like consistency, without any evident fibers or structures. Many thousands of red-brown seeds were densely concentrated in bands in numerous, vertically oriented radii. The fruit was so oily that oil pools formed in hollows where the flesh had been removed to the base, and we had difficulty drying the fruit after collection. The flesh had an extraordinarily strong, penetrating odor like slightly rotten coconut. It tasted sharp, acidic, and somewhat like coconut.

This strange fruit is not an anachronism in the profound way that avocado may be an anachronism. Unlike gomphotheres, elephants and rhinos are not extinct. But they are locally extinct; they have been extirpated throughout the remaining range of *Rafflesia*. If one can expect to encounter ecological anachronisms in many habitats of the New World, *locally* anachronistic fruits are evident just about everywhere and increasing every day.

Local Losses

If you share habitat in North America with honey locust, Kentucky coffee tree, or mesquite, you live among ecological anachronisms. If pawpaw, persimmon, or osage orange are in the neighborhood, you should be able to make out the ghosts of their extinct partners too. The same holds for anachronistic burs, like devil's claw and cocklebur, and for the anachronistic thorns of mesquite, honey locust, and osage orange. Anywhere in the Northern Hemisphere, you might scare up a ghost in a hawthorn hedge or among the spiny leaves of a holly tree. If you live in a temperate city you can commune with Mesozoic ghosts in the shade of a ginkgo tree. Finally, neotropical anachronisms such as avocado and papaya are no more difficult to locate than a well-stocked grocery store.

Even where anachronisms are not in evidence, one can find bereaved plants of another sort. These might be called the *ecological widows*. Like *Rafflesia*, their pollen, fruit, or armament is intended for animals not extinct but extirpated locally or regionally. If the riddle of the rotting fruit challenges you in your own backyard, suspect a missing mutualist. If a plant with showy flowers bears scant fruit, remember the forgotten pollinators. If deer nibble leaves between widely spaced thorns, remember the bulky, big-mouthed mammals who once towered over them.

Even in the wilds of New Mexico, ecological widows are sadly in evidence. No one has written about them as such, but the clues are compelling. Consider the case of the little yellow "tomato."[35] That's the name I prefer for *Solanum elaegnifolium*, rather than the traditional name of sil-

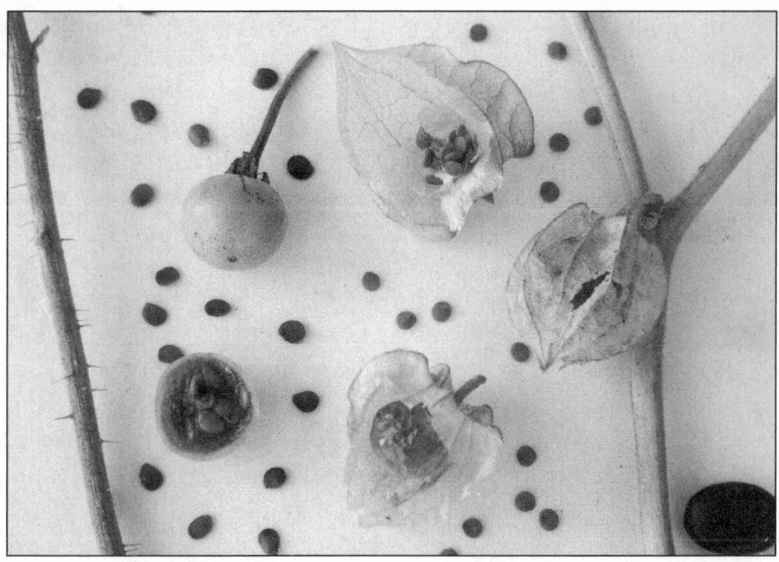

A range of adaptations in the tomato family. The tomato-like yellow fruits of *Solanum elaegnifolium* are contrasted here with the dry seed packages of a species of closely related *Physalis*. Only the pulpy fruit needs stems armed with prickles to repel slugs and snails. Honey locust seed (1 cm) for scale.

verleaf nightshade. While conducting gourd counts in the floodplain of the Gila River, I got to wondering about the little yellow tomatoes rotting or desiccating on upright stems amid the rotting or desiccating gourds on prostrate vines. Who—and where—were their dispersers?

This perennial, deep-rooted plant belongs to the tomato and potato family, Solanaceae, which produces some of the most potent alkaloids of the entire plant realm. I was not therefore tempted to chew and spit the fruit as I usually do. "One or two will probably kill you," warned my neighbor Allen Campbell. Apparently the critters know that too. The little fruits retain their moist pulp for several months; by winter some have browned and succumbed to rot while others retain their bold yellow but are dry inside. Almost all remain aloft—seeds trapped in a prison of the plant's own making—even into the following summer, as a new crop of purple flowers blooms.[36]

A search in the scientific literature confirms what field experience suggests. *S. elaegnifolium* and its sister species *S. carolinense* inhabit disturbed sites and floodplains—the former in the desert Southwest and the latter in the eastern half of the country, overlapping in the midwestern states. The usual contingent of mammal and bird frugivores do not regard either as food, and its glycoalkaloids can be lethal to livestock. Indeed, a 1997 paper presents the fruit of *S. carolinense* as a bit of a mystery, and phylogenetic inertia is the only causal explanation offered.[37] To my way of thinking—having been reprogrammed by Dan Janzen and Paul Martin—phylogenetic inertia is no explanation at all. Let us then search for and evaluate the best adaptive story. Was there once a dispersal mutualist for the ancestors of these plants along the Gila River? If so, is that disperser extinct, or has it merely been extirpated from this locale?

To begin, we must discern who the plant is trying to repel, as well as who it is trying to woo. Why is the fruit defended by intense alkaloids? Simple observation suggests that the poisons in the pulp of *S. elaegnifolium* do a superb job repelling insects. All the fruits I examined were insect free. But then, so, too, have been all the homegrown tomatoes (*S. lycopersicum*) I have enjoyed. For edible tomatoes, extreme acidity in the pulp may be insecticide enough. Slugs will devour tomatoes that touch the ground, however. It is thus noteworthy that stems of *S. elaegnifolium* have prickles that break away at the merest touch and painfully embed in the flesh of the toucher. Slug deterrents they are.

The little yellow tomato is unquestionably overdefended for dispersal agents locally available. Evolution should have left a window of palatability for at least some winged or legged vector. I think I know who that vector is intended to be. Perhaps some year I will have a chance to test the idea. All I need do is deliver the ripe fruits to a turtle or tortoise accustomed to wild cuisine.

Box turtles and desert or gopher tortoises are found within the wide geographic range of the two sister species of *Solanum*. Neither lives anywhere near my New Mexico home, however. Not surprisingly, then, the yellow tomatoes in my region grow either in the floodplain of the Gila River or on a terrace just above it, where the floodplain used to be. Floods may well be the only reliable vector dispersing the seeds. Dispersal may not be vitally important to this plant, however, at least over the short term. The stems and leaves die back every year, but the taproot is deep and perennial. Considered weeds, these *Solanum* species are difficult to eradicate once established. An individual plant more than maintains; it propagates by root runners. Perennial, cloning roots are compensatory life history traits for plants bereft of dispersal partners.

Several other traits support this turtle hypothesis. First, the little yellow tomatoes are small enough for a turtle or tortoise to consume. They hang down from a plant that grows no higher than my knee and that does not shed ripe fruits. Most fruits are thus well within the reach of turtles—and beyond the reach of slugs and snails. All would have been accessible to the *Geochelone* tortoises that went extinct in North America thirteen thousand years ago.

A casual experiment on my porch indicated that our resident mouse will not gnaw into a yellow tomato. Nor will it extract seeds exposed on a freshly halved, moist fruit. But when I extracted and washed the seeds and left them on the ledge, all that remained the following morning were empty seed coats. The mouse had shelled the seeds and eaten the contents on the spot.

The library offers additional clues. To begin, one can learn that the ornate box turtle (whose shell is shown in the photograph that opened this chapter) is fond of garden tomatoes. *Terrapene ornata* also favors another member of the tomato family: ground cherry *(Physalis)*.[38] Its close kin the eastern box turtle *(Terrapene carolina)* does a superb job of dispersing the yellow fruit of the mayapple, which is found in mature forests of eastern North America.[39] The mayapple, *Podophyllum peltatum* (family Berberidaceae), grows to about the same height as the yellow tomato plant, and it, too, has perennial roots that spread vegetatively. Mayapple bears the third largest fleshy fruit native to temperate North America and edible to humans—after the pawpaw and persimmon. Like the tomato, a mayapple fruit hangs down, and it is obscured from above by huge, umbrella-like leaves. Passage through a turtle is known to increase the germination success of mayapple seeds.[40] In the Galápagos Islands, native tomatoes will not germinate at all unless they have passed through the gut of a giant tortoise.[41]

Both the mayapple and the *Solanum* that stimulated my quest bear yellow fruit. Fruits attractive to birds are typically red, dark blue, or black. Fruits attractive to mammals are often green or brown and make their presence known by odor more than color. Box turtles possess an enriched optical sensitivity for the yellow-orange part of the spectrum. There is also the matter of toxins. Reptiles are repelled by tannins in their food, but they far surpass mammals in their tolerance for alkaloids. Indeed, they cannot taste bitterness, which is often the sign of an alkaloid. When their food is artificially spiked with the alkaloid strychnine, captive turtles will dine till they drop.[42]

If the ecological ghost is in fact a turtle or tortoise, how did the little yellow tomato arrive and persist near my New Mexico home? A seed may have been carried on the tire of a car or boot of a tourist rather than in the gut of a turtle. The only populations of this plant that I have encountered have been within a kilometer of a road, and roads are notoriously easy pathways for invasions of exotic species. Climate change is another possibility; the turtles may have been driven to lower elevations but the plant hung on. There is one other reasonable explanation: the fruit is an ecological widow.[43]

European invaders have never had a substantial presence along the upper reaches of this river, although our cattle have. These canyons and mountains were the last stronghold of the Apaches, who were forced to surrender little more than a century ago. The most intense human occupation, rather, is thought to have occurred a thousand years before my culture usurped the land. The terraces and caves along this stretch of the Gila River yield plentiful artifacts of an agricultural people: pottery chips, obsidian flecks, cliff dwellings, pictographs, even mummified husks of corn. Turtles and tortoises would have been vulnerable to overharvesting, especially if they depended on the narrow floodplain. Expansive regions of the eastern woodlands of North America are barren of box turtles, and the reason is thought to be overexploitation by indigenous peoples, especially those for whom tortoise shells were ceremonial objects.[44]

Humans are humans whenever and wherever we may be. We take what is easy to take, especially when we enter a new land.[45] The wisest cultures then adapt toward sustainability with the biodiversity that remains. Excellence in land stewardship, however, may have little overlap with excellence in defending territory against land-hungry peoples. Whatever role one or another group of Paleo-Indians may have played in the end-Pleistocene extinction of big game throughout the Americas, and in local extirpations of smaller creatures in the vicinity of settlements, that degree of damage is nothing like what my own culture is poised to carry out.

"Events of the last forty thousand years may be viewed as the preamble to where we stand today," Paul Martin has written. "What happened to the mammoths and to the diprotodons of Australia even earlier was the beginning of a cascade of extinction that swept away thousands of species of land birds and other animals, and on some oceanic islands, even the native land snails and trees. . . . These losses are soon followed by the much more familiar extinctions of historic time, the dodo of Mauritius, the solitaire of Réunion, Stellar's sea cow on the Commander Islands, the great auk, Labrador duck, Carolina parakeet, and Tasmanian wolf, to begin a litany of the names from the Doomsday Book. Today we find ourselves on the brink of an extinction spasm, about to lose one quarter of the 250,000 higher plant species on earth. . . . The biogeographic pathologies of the last forty thousand years, a mere moment in earth history, bear hallmarks of impending mass extinction."[46]

Ecological anachronisms and widows are at the ready to teach us about consequences. Shall we listen?

Devolution

"*Maclura* is on the verge of becoming a clonal species," Peter Del Tredici declared during an interview at Arnold Arboretum. We had been talking about his research on ginkgo, and the conversation broadened to other anachronistic fruits. He explained that osage orange, genus *Maclura,* can reproduce asexually by producing fully formed fruits containing seeds that are clones of the parent. A female osage orange tree may not therefore require a male to be anywhere upwind (the flowers are wind pollinated) in order to produce viable seeds inside fruits that are attractive—or were attractive—to megafaunal dispersers.

"*Maclura* also root suckers like crazy," Del Tredici told me, "and appears to be in the process of losing its sexuality."

"I guess that makes sense, if nobody is dispersing its fruit," I responded.

"Well, I don't really know how quickly evolution can take care of something like that," he corrected. "Ten thousand years doesn't seem that long. It's hard for me to believe that a tree that can sucker and that has a lifespan in excess of a hundred years for a single stem is going to be able to shift its sexuality that quickly. Ten thousand years is not that many generations for such a plant."

Recall that Peter Del Tredici is accustomed to thinking about consequences for a plant that may have lost its target dispersers sixty-five million years ago. The passage of ten thousand years would not alter the genome of the ginkgo lineage. The single species of ginkgo alive today may, in fact, be sixty million, even a hundred million, years old. It has changed that little despite terrible turnovers of animal mutualists.

Ginkgo, a gymnosperm, is the extreme. Nevertheless, perennial plants in general evolve at a much slower pace than do the animals that disperse them. Fossil evidence suggests that the mean duration of angiosperm tree species is thirty-eight million years. For shrubs the figure is twenty-seven million. In contrast, mammal species come and go in about one-half to four million years. Bird species are more ephemeral; most have a longevity at the low end of the mammal scale. Carlos Herrera reviewed this discontinuity between plant and animal evolution in a 1985 paper, "Determinants of Plant-Animal Coevolution." He noted that plant structures for dispersal persist over long geological periods and without noticeable change.

Fruit forms may be even more enduring than species identities. We can see this stability even without consulting the fossil record. Plant genera whose species are widely disjunct in geographic range (including those on opposite sides of the globe) maintain a surprising constancy in fruit traits such as color, size, and nutritional value of pulp. Herrera draws a bold conclusion: *all* vertebrate-dispersed fruits are anachronistic. He supports his claim by noting that none of the animal species that were interacting with plants when the fruits first evolved are alive today. By this standard, even the gomphotheres weren't the real thing for a Mexican avocado. Only an early Cenozoic brontothere would do.

Herrera's claim is an unhelpful extension of the term *anachronism*. Plants do not interact with taxonomic species; they may not even interact with taxonomic families. Rather, they interact with mouths and guts and peripatetic tendencies. That the gomphotheres or brontotheres are gone is not the issue. What matters is that animals with big mouths, gentle interiors, and serviceable legs have disappeared from the avocado's homeland. These crucial traits were possessed by many of the missing megafauna. Fruits that evince what Paul Martin calls the megafaunal dispersal syndrome were shaped by interactions with big animals long ago. The hind- or fore-processing distinction may make a difference to many such plants, especially the biggest-seeded or sweetest kinds. But brontothere or gomphothere is a detail that does not register. To this way of thinking, perhaps, an American avocado tree growing in an orchard on the periphery of an elephant reserve in Cameroon is not anachronistic, because African elephants occasionally raid the orchard and later defecate viable avocado seeds.[47]

Dan Janzen coined the term *ecological fitting* (in contrast to the traditional understanding of *evolutionary fitting*) to apply to all such instances in which a lineage outlives the biological companions that gave it form, or when a lineage finds itself in a new habitat or climatic regime and must make do if it is to persist. Anachronisms that haven't sputtered out of existence are examples of ecological fitting. Janzen wrote: "A major part of the earth's surface may be occupied largely by organisms that are rich in ecological interactions and have virtually no detailed evolutionary history with one another. In such a view, questions such as 'What are the selective pressures maintaining such and such a set of traits of an organism?' would need to be severely augmented by 'What might have been the selective pressures that invented such a set of traits?' also 'What blocks their decay?' and 'What are the ecological processes that lead to such and such a kind of ecological fit?'"[48]

Decay may be the future of many anachronistic traits. Devolution that transforms a lineage at the level of the genes goes beyond ecological fitting. This is evolutionary fitting. The lineage is pressed to change genetically to fit into a new matrix of biological companions and climate. If traits are "evolutionarily frozen," however, ecological fitting at the individual and population levels will be the sole recourse. In some habitats a species will be common; in others it will be uncommon but still able to breed; yet others may contain strays (or long-lived clones that had established during favorable times), but there will be little breeding success. Extreme anachronisms are noteworthy in that nonbreeding populations may be the only ecological fit available to some of them in the wild.

Unambiguous evidence that plants can, in fact, reverse the process of evolution—that they can devolve an overbuilt or suddenly maladaptive trait—is found on oceanic islands that lack large, indeed any, mammals. *Solanum* species, like the little yellow tomatoes along the Gila River, lose their stem prickles once they establish on mammal-free (or perhaps slug-free) oceanic islands. Burs lose their stickers.[49] Flies may lose their wings. Caves anywhere in the world are cauldrons of devolution, as fish and arthropods isolated in the dark lose sight and eventually their very eyes.

Pressure for devolving large fruits and seeds is evident today in the neotropics. A nutmeg, *Virola surinamensis*, establishes best from larger seeds, but the birds that are its only effective dispersers have been found to favor nutmeg fruits with seeds at the small end of the spectrum of variation and to avoid the large.[50] Unless a balance of trade-offs has already been attained, devolution of seed size should be expected, especially since this trait is already present in the population. What might be expected, however, when more drastic change is called for? What shift might plausibly occur in a lineage, such as osage orange or Kentucky coffee tree, that produces fruits too big and too poisonous to woo anybody with whom it currently cohabits? Can an overbuilt trait be more than just trimmed? Can an anachronism be retooled?

What fate might we imagine for that little yellow tomato along the Gila River? Might this population of *Solanum elaegnifolium* transform into something like the airy lantern of its sister genus *Physalis*? How much energy and resources would the now overbuilt tomato save by that kind of devolution? Indeed, what selective forces, or changes in selective forces, may have pressed the ancestors of *Physalis* to dispense with pulp in the rudimentary berry at the heart of its lantern? After all, ground cherry and tomatillo belong to this genus too, yet within each of their lanterns can be found a berry still plump with pulp—and plumped up even more by generations of humans selecting the best varieties for food. Would the little tomato come to resemble *Physalis* in another way too: shedding stem prickles? Might prickles vanish when the fruit had cast away pulp to the point that slugs no longer found it attractive?[51] The two taxa pictured in the photograph on page 215, by the way, were collected from sites just a few hundred meters apart.

If such an imaginative exercise is pointless, then Dan Janzen has dabbled in pointless speculation too. He has suggested that in the period since the megafaunal extinction, a tall neotropical tree distantly related to Asian mango may have shifted toward longer retention of ripe fruit in the canopy. Such a shift could have occurred if, as he suspects, bats and monkeys disperse the stony pit of this yellow, plum-size fruit more effectively now than deer. None of these animals will swallow a pit, but monkeys and bats eat far more of these fruits than will deer. Stripping a fruit of its pulp while sitting in the branch of the parent tree does the seed no good, but assistance is provided whenever monkeys and bats carry fruits to sites where the flesh can be eaten in safety.[52]

For the Chihuahuan desert, Janzen augurs that wind-dispersed cactus seeds might evolve in the future, even though the cactus family relies today entirely on consumption of fleshy fruits for propagation by seed. Whatever evolutionary changes might be in store, annuals are likely to get there first, Janzen suggests, because these plants have only a one-year lapse between generations.[53] Meanwhile, the slow-growing, long-lived members of the cactus clan will continue to rely on vegetative, clonal means of reproduction—exceptionally well developed in the burlike adaptations of the stems of jumping cholla.[54]

By far the most provocative example of post-Pleistocene adaptation is in the honey locust genus, where devolution is not relegated to some speculative future. I maintain that it has happened already, and that both the original and the devolved fruit types coexist today. An extraordinary kind of devolution seems to have occurred among the fruits of genus *Gleditsia*—not once but twice and on two continents.[55] If the impetus for such change has been, as I believe, the end-Pleistocene extinction of large hindgut processors in North America and their extinction or widespread extirpation in Asia, then the fossil record should support this claim. I therefore predict that there will be no fossil evidence of the devolved fruit type before each continent was stripped of hindgut megafauna.[56]

The elder and devolved fruit types in eastern North America can be seen in the accompanying photograph. The ancestral fruit would be the long, curving pod of honey locust, *Gleditsia triacanthos*, which regularly contains twenty or more seeds.[57] This is the species we see along urban streets and parking lots, only females of which, however, bear fruit. The devolved lineage, whose fruit contains a single seed, is the diminutive pod of swamp locust, *Gleditsia aquatica*. An intermediate form occurs rarely where the two ranges overlap; it is flat like a swamp locust pod but has the multiple

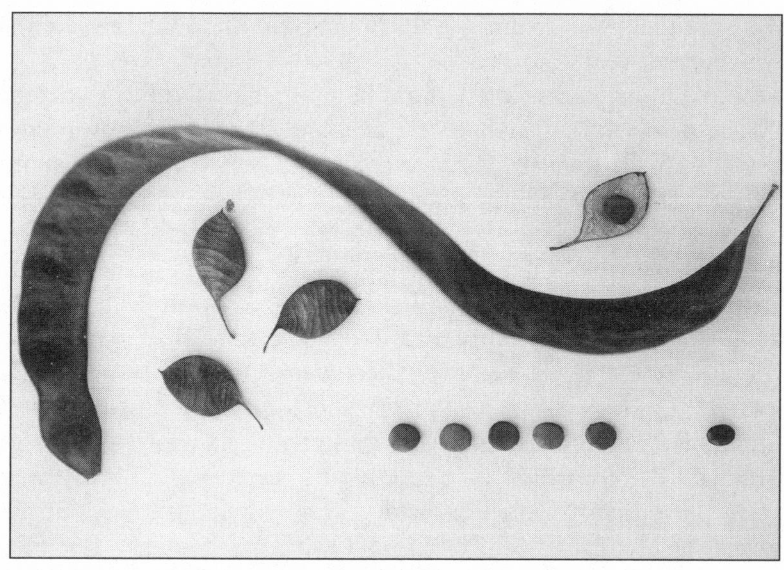

Devolution in the legume family. The diminutive pod of swamp locust contrasts with the curving fruit of honey locust. These sister species of genus *Gleditsia* may have diverged since the end of the Pleistocene. If so, swamp locust would have devolved from the ancestral pod shaped by megafauna.

seeds of a honey locust and is about a third the length of the latter. This intermediate was once considered a separate species, *Gleditsia texana,* but is now officially a hybrid of *G. triacanthos* and *G. aquatica.*

Several features of swamp locust suggest adaptive reasons for the shift away from the honey locust standard. The single seed is most apparent. If a lineage can expect its fruit to be crushed in the molar mill of a large mammal, with seeds defecated over the course of days or weeks and perhaps further isolated by dung beetles, then it doesn't matter how many seeds are crammed into a pod. When such helpers vanish, however, sibling seeds trapped together will be forced to compete—if indeed they can sprout from an unopened pod. The post-Pleistocene has thus pressed this lineage to isolate seeds, one per pod.

What sort of pod? The photograph cannot convey the difference in the thickness of pod walls. A swamp locust seed is packaged between valves no more robust than that of the papery shell of black locust, *Robinia pseudoacacia,* which is an enigmatic North American legume in its own right. Unlike those of *Robinia,* however, the pods of swamp locust do not open automatically. Environmental abuse of some sort is necessary to liberate the seed. Still, far less abuse is required to sunder the pods of swamp locust than of honey locust. I have seen year-old honey locust pods in New

York City intact on the ground, seeds imprisoned and ungerminated, even after the current year's crop has joined the collection.

Another difference between the sibling species is found within the pods. Ripe honey locust pods have pistachio green, sugary pulp. Swamp locust pods are as dry as *Robinia*. Swamp locust has evidently caught on to the fact that the megafauna are gone and has stopped laying out snack food. Pulp devolution has also affected wild populations of honey locust. Sugar content of ripe honey locust pods in the wild ranges from thirteen to thirty-one percent.[58] Although I have yet to encounter a wild honey locust, I suspect that the pods I have tasted along urban streets are the leanest of the lot. They also may be more tannic than wild strains. Just as sugar content may devolve, so, too, the species may be putting less effort into immobilizing tannins with which it defends growing pods.

The small, flat pods of swamp locust will more easily take to the wind than will honey locust pods. Both kinds are buoyant, but swamp locust can expect to be flushed into more nooks and crannies in a floodplain.[59] There is, after all, a reason for the common name *swamp locust* and the species name *aquatica*. In the wild it is found almost exclusively in floodplains. Honey locust probably was too, before livestock from Europe initiated a new episode of range expansion. Only swamp locust grows in the wettest landscapes, however, including swamps.[60] Both grow well in uplands, but somebody has to get them there.

Pods and pulp are not the only differences in the two fruits. Swamp locust and honey locust bear distinctive seeds. The seeds weigh about the same, but that of swamp locust is a squashed version of its presumed ancestor. Shaped like a pancake, the seed of swamp locust also has a thinner seed coat (0.87 mm compared with 1.18 mm for honey locust).[61] Millions of years ago, genus *Gleditsia* equipped its seeds with coats tough enough to withstand grinding teeth and churning gastric juices. Absent such punishment, germination may be held back for three or more years, so an overbuilt seed coat is now a liability.[62] Swamp locust has made a wise move in thinning its protective shield. Why hasn't honey locust followed suit?

In what seems a contradiction, swamp locust seeds are invulnerable to predation by bruchid beetles, even though their seed coats are thinner.[63] How is this possible? The answer is that swamp locust is not merely thin; it is too thin. Its flattened seeds are too flat to house the plump larva of a bruchid beetle. Flatness is thus a great adaptation. Why didn't *Gleditsia* switch to pancakes for seeds a long time ago? Malcolm Coe has found that among African pod-bearing *Acacia* trees, the flattened seeds of wind-dispersed pods are less vulnerable to bruchid attack than are the thicker seeds of animal-dispersed pods. He suggests that the seeds in animal-dispersed pods evolved sphericity primarily to resist destruction in molar mills.[64]

The same devolution of fruit form has happened independently in Asia. *Gleditsia microphylla* joins *G. aquatica* as the only species with pulpless pods in this worldwide genus of thirteen taxa. As well, all other species bear at least ten seeds per pod; *microphylla* has one to four. Although both species of the North American pair produce the same leaf form, the Asian pair demonstrates a dichotomy in sync with pod devolution. The robust pods of *G. fera* (the species used to be called *G. macrophylla,* "big leaf") are borne on branches with exceptionally large leaflets, whereas the diminutive pods of *G. microphylla* hang amid a lacework of leaves.

A cladistic analysis, based on differences in chloroplast DNA among the thirteen species of genus *Gleditsia,* posits a somewhat greater taxonomic distance between the Chinese pair than between the North American pair.[65] Nevertheless, *G. fera* is the closest relative of *G. microphylla.* As to the deeper, intercontinental divergences, Andrew Schnabel and Jonathan Wendel conclude that genetic separation between North American and Chinese stock occurred four million years ago. Ergo, swamp locust must have devolved from honey locust much more recently than that. Indeed, the authors found very little divergence in chloroplast DNA between swamp and honey locust, indicating "a recent separation" of these two lineages.

Pods, pulp, and seeds may not be the only traits of *Gleditsia* that have devolved since the megafauna vanished. Of the thirteen *Gleditsia* species (none of which are native to Africa, by the way), six have both thorned and thornless variants.[66] Had not honey locust spun off a thornless mutant, there would be few honey locust trees lining streets and parking lots. The thorns are far too dangerous. Nevertheless, in the chromosomes of these sorry mutants lurks a genetic memory that reactivates with a little provocation. Take pruning shears to—or unleash an elephant upon—a thornless cultivar, and thorns may magically reappear.

In addition, the thornless variant does not breed true. This suggests a recent and still unsettled adaptation. On average, one out of five seeds produced by thornless females pollinated by thornless males will yield thorny offspring.[67] I have observed ferocious spines on two seedlings growing in a bed of ivy in my apartment complex, a few meters beyond the canopy of an adult, thornless female and surrounded by a scattered population of thornless street trees.

It will be a sad day for honey locust if swamp locust ever throws up a thornless mutant in a place where botanists might notice. Swamp locust, with its dainty pods, will then be the urban tree of choice. That will be a sad day, too, for ghost watchers. A thornless swamp locust would have lost the last traces of millions of years of evolution in the company of giants.

SEEDS OF ANACHRONISTIC FRUITS IN NORTH AMERICA. From biggest to smallest: gymnocladus, pawpaw, honey locust, persimmon, mesquite, two genera of desert gourd, devil's claw, maclura, and opuntia cactus.

10

The Great Work

LONG BEFORE HUMANS BECAME HORTICULTURALISTS, we were seed dispersers. Pawpaw, whether planted deliberately or sprouted from middens, is associated with sites of former Native American occupation. So, too, is Kentucky coffee tree; its big, spherical seeds took a fine polish, prompting Native Americans of the eastern forests to collect and carry the seeds for gaming sports.[1] Similarly, Gary Paul Nabhan has written, "For the last 9,000 years, gourds in arid America have gotten around in another way—propelled by the hand of man. Long before Fernando Valenzuela signed up with the Los Angeles Dodgers, his Indian ancestors were pitching and ducking gourds the size of baseballs. Wild gourds appear preadapted to the human hand."[2] *Cucurbita foetidissima* would have been valued by practical adults as well as cavorting children. The thick-rinded gourds make serviceable containers. And washed clean of its bitter pulp, the seeds are as edible and nutritious as those of domesticated squashes. Tasty, too, I can attest. I washed and roasted desert gourd seeds, along with a batch of cultivated squash seeds, and could not detect a difference.

Further south, Paleo-Indians valued desert gourd's relative, *Cucurbita pepo,* for pulp as well as seeds. For at least ten thousand years, the Indians bred bitterness out of and sweetness into that little gourd. Today we marvel at the diverse, souped-up results: the summer squashes, zucchinis, and pumpkins.

Five years after Dan Janzen speculated on the role that humans may have played in the persistence of jicaro gourds, *Crescentia alata,* in Central America, Gary Paul Nabhan charged ahead with the idea.[3] "If one looks at the present-day distribution of *Crescentia* species," Nabhan wrote, "neither water dispersal nor horse dispersal appear to have had as much influence as human dispersal."[4] Nabhan assembled a list of 112 neotropical species of fruits "presumably eaten by now-extinct megafauna."[5] Of these, he concluded that at least 98 "persisted as culturally utilized plants." Human appeal is thus a life history trait of anachronistic fruits that compensated for the loss of megafauna. In the Western Hemisphere, four out of five tropical and subtropical species ecologically stranded by the loss of their mutualists may have been rescued, in part, by human attention. In the case of *Cucurbita* and many other genera, humans did more than compensate; we nurtured the lineage into newfound splendor.

Nabhan speculated that the close match between fruits expressing the megafaunal dispersal syndrome and those appealing to humans was no coincidence. "Megafaunal selection of fruit qualities preadapted them to human use in ways that made unnecessary much counter-selection by neotropical cultures." When I enjoy an avocado, I should be grateful not only to hundreds of generations of human horticulturalists but also to hundreds of thousands of generations of mouths and guts a great deal larger than my own.

Gary Paul Nabhan got on the track of anachronisms while a student at the University of Arizona. He dedicated his megafaunal paper to "my mentor and friend Paul S. Martin." Martin, in turn, recalls that Nabhan was "the first to get me thinking about humans as proxies for megaherbivores in fruit dispersal after megafaunal extinctions. That led to the idea that, in the brief time between colonization and extinction, the large animals revealed much of what they knew about edible plants to the first people."[6]

Gardening for Diversity

Thirteen thousand years ago, the responsibility for dispersing edible, large-fruited plants in the Americas fell to humans. The present demands another profound passage in stewardship. We must expand our circle of care to encompass those plants from whom we can expect no reward. Keeping such species alive promises us no delicious fruit or useful artifact, no lacy-leafed tree to beautify our cities. We will keep these plants alive solely because we choose to honor the past and carry forward the richness of life on Earth. The more animal mutualists we extinguish now, the more plants we will be obliged to partner when our species awakens to a new ethos, which Thomas Berry depicts as "a mutually enhancing human-Earth relationship."[7]

Is our species capable of such selfless service? Will we have the knowledge and the verve to succeed? When will we put an end to this saga of loss upon loss and resolve to undertake what Berry calls "the Great Work" of ecological and evolutionary restoration?

On the island of Mauritius in the Indian Ocean, humans are selflessly serving the tambalacoque tree.[8] We are stripping the (to us) inedible pulp from large stony pits, a service the dodo dependably provided until three hundred years ago. Stripping pulp to prevent seed rot is not all we must do. Seedlings returned to the "wild" must be protected from fast-growing interlopers—the nonnative plants we intentionally set loose on the island and now cannot eradicate. We may also need to fence out browsers, notably deer—another such interloper. The ecosystem of Mauritius has been so thoroughly scrambled that becoming gardeners for diversity is no easy task.

Even more arduous is the kind of gardening demanded of human stewards in one small natural area of northern Florida. Torreya State Park is perhaps the most depressing preserve I have ever visited. On the one hand, its botanical richness is a glimpse into the past and a splendor to behold. On the other hand, even this richness is not sufficient to ensure the continuance of a very special tree.

Torreya State Park boasts elegant beeches and soaring pines. The dense canopy nurtures the kind of shade-tolerant undergrowth that stuns us with its delicacy. A postage-stamp preserve, the park and the adjacent Nature Conservancy property are surrounded by pine plantations managed for pulp, by cattle pastures managed for meat, and by the tangle of fierce plants that take hold wherever humans take up residence. Inside the preserve are troubles too. *Torreya taxifolia,* the tree for whom this park is named, is the world's most endangered conifer. Worse, those who have assumed the role of protector are resigned to the idea that this tree is almost certainly doomed to extinction, no matter what they may do.

Here is where anachronism enters the story. The genus *Torreya* (yew family, Taxaceae) was once distributed throughout the Northern Hemisphere. Range fragmentation has produced distinct species in eastern China, northern Florida, and patches of the Coast Range and the Sierras of California. The Florida species of torreya has a geographic range today restricted to the cool ravines along the east side—just the east side—of the Apalachicola River in northern Florida and a bit upstream into southern Georgia. The Apalachicola deserves our gratitude for having welcomed into its moderate climate and rich soils all the trees and forbs that now enrich the Cove Hardwood forest of Great Smoky Mountains National Park six hundred kilometers to the north. The Apalachicola was a refuge for these plants during the peak of glacial advance some twenty thousand years ago. After the ice retreated, the plants hitched rides from wind and animals in order to return home. Torreya was left behind.

The three botanists I queried about the Florida torreya all agreed that the tree's troubles may well have begun for want of a disperser.[9] Extinction or regional extirpation of an animal partner (or partners) may be the ultimate cause of the tree's current distress. Torreya is probably not suited for the warmth and humidity of the Apalachicola region today. It wants to head north, but it hasn't found a vehicle.

FLESHY SEEDS OF THE WORLD'S MOST ENDANGERED CONIFER. This jar is the only way for a visitor to experience such seeds in Florida's Torreya State Park. Today the seeds of *Torreya taxifolia* are too precious to embalm. At most a single seed matures in the park each year.

Nevertheless, the proximate cause of the tree's imminent extinction is disease. Some thirty different pathogens, notably a devastating fungal blight, have pounced on this evergreen conifer since the 1950s. Like the American chestnut that was destroyed by (an imported) blight during the early years of the twentieth century, torreya keeps sending forth sprouts from the base of dead or dying stems, only to repeat the cycle of death and futile regrowth. Unlike the ginkgo, which bears strikingly similar pulp-covered seeds, torreya has no lignotuber (or Chinese monks) to rely on. Energy stored in its roots eventually gives out, as replacement stems in the shady forest consume more photosynthate than they can return to the roots before their demise.

That Florida torreya may be haunted by the ghosts of extinct dispersers is suggested by a host of evidential threads. First, torreya fruit (technically, a diaspore) is distasteful to many mammals who normally consume fruits.[10] The pulp has a high terpene content, and it leaves a sticky residue on one's skin. Squirrels treat the fruit as they treat ginkgo fruit in New York City parks: they steal the seeds. Whether scatterhoarding then ensues and whether that interaction is helpful to torreya are unknown. Once seed production falls below a threshold of abundance for whatever reason, decline may be exacerbated by rarity. Squirrels likely consume rather than store any rare seed they come upon. As the population diminishes, receptive ova are bathed by winds that no longer bring them pollen.

In contrast to the plight of the Florida species, California torreya is maintaining its population, as are the several species of Chinese torreya—all of which bear identical propagules. One important difference is that the California habitat is much drier than the Florida enclave. Another thread of evidence is that, before the blight took hold, Florida torreya was planted as a landscape tree and novelty on a number of private properties near Wilmington, North Carolina, most notably the Biltmore Estate. The largest and healthiest torreyas are now living in North Carolina, not Florida. All is not well even in the north. Although the Biltmore trees have not yet suffered blight, some surrounding plantings on private properties have weakened or died of disease.

Florida torreya is so endangered that experimental use of seeds or cuttings is strictly controlled. How I would love to see humans step in for the absent disperser—whoever that disperser may have been. A megafaunal ghost perhaps? Or maybe a large extinct or extirpated tortoise? How I would rejoice if plantings were undertaken on some cool slope of the Smoky Mountains. Such is not, alas, how things are done with endangered species—the lone exception being the recent return of the California condor to its Pleistocene home near the Grand Canyon. Native territory is regarded as the last best place to be. It is sacred. That may be sound practice in some cases—but not all. In a study of endangered species published in 2000, Rob Channell and Mark Lomolino concluded that "most species examined persist in the periphery of their historical geographic ranges." The implication is that the last place a troubled species is found may not, in fact, be the best place to assist its recovery.

Transplantation is an uncommon and controversial technique for biodiversity conservation today. But after the greenhouse effect has ratcheted up temperatures and rerouted rainfall, it surely will become the norm. If gardening a few local patches of endangered plants is tough today, it's going to get a lot tougher when, like it or not, we become gardeners of the planet. Helping plants track climate change from one patch of forest to another will be a routine tactic for conserving biodiversity decades hence. Is it too early to begin now with Florida torreya?

Resurrection

"I want to leap ahead of the understanding that we finally accept extinction as a phenomenon—in fact, we're enveloped by it—to the possibility of doing something about extinction. The impetus for this, in the past, came from our yearnings for the reconstruction or the resurrection of the sort of wilderness that Lewis and Clark had experienced while exploring the American West. But the redemptive aspect of what I propose is radical, and I'm not sure it's going to go beyond this audience."

Paul Martin was speaking to a small group at the civic center auditorium of Hot Springs, South Dakota. It is late June 1999. He and I traveled to this town at the southern edge of the Black Hills in order to participate in the twenty-fifth anniversary celebration of the discovery of the greatest collection of intact mammoth skeletons in all of North America. The huge, partially excavated pit with in situ bones at the "Mammoth Site," as it is called, is protected by a soaring edifice. From a distance the building looks like a church. Paul calls the site "a paleoecological cathedral."

Paul had just finished the introductory section of his slide presentation, in which he reviewed the timing and magnitude of the "extinction of the massive" at the end of the Pleistocene. Earlier, the ghost of Europe's esteemed anatomist Georges Cuvier (namesake of our avocado-eating gomphothere, *Cuvieronius*) had addressed the gathering.[11] Cuvier delivered a eulogy to the great mammoths, explaining how the discovery of fossil bones—especially mammoth molars—had compelled him in 1796 to conclude that not all species created on this earth were still alive: "What has become of these two enormous animals [mammoth and mastodon], the remains of which are found everywhere on earth and of which perhaps none still exist; beings whose place has been filled by those that exist today, which will perhaps one day find themselves likewise destroyed and replaced by others? Why, lastly, does one find no petrified human bone? All these facts seem to me to prove the existence of a world previous to ours, destroyed by some kind of catastrophe. Modest anatomy will be content with the honor of having opened up this new highway to the genius who will dare to follow it."[12]

Before Cuvier made his bold pronouncement, Western humans had no awareness of extinction. It was thought that surely somewhere in the unexplored regions of Earth the great beasts still survived in the flesh. We were ignorant, too, of the long reach of Earth history, of the hundreds of millions of years of animal grandeur that had preceded the birth of genus *Homo*. It would be another half century before Charles Darwin would astound the world with his *Origin of Species*. And, of course, neither Cuvier nor Darwin had an inkling of the extinctions that lay ahead.

Thomas Jefferson's press secretary had preceded Cuvier at the podium.[13] He delivered a message in behalf of the third president of the United States, who extended an apology for not being able to deliver, in person, his eulogy to the mammoths. Jefferson had launched the Lewis and Clark expedition in 1804. He sold Congress on the idea because of the chance it would offer to find a trade route from the Mississippi River to the Pacific

THE MAMMOTH SITE IN HOT SPRINGS, SOUTH DAKOTA. In the company of North America's greatest aggregation of intact mammoth skeletons, Paul Martin in 1999 launched his idea to "bring back the elephants." Courtesy of the Mammoth Site.

Coast. But he also dearly wished to prove Cuvier wrong. In private conversation, he directed Meriwether Lewis to keep a keen eye out for signs of mammoths in the flesh and perhaps even the strange creature whose partial skeleton Jefferson himself had described in a scientific publication a few years earlier. That creature was a ground sloth, which Jefferson misinterpreted as a carnivore because of its prodigious claws. Fifteen years before the expedition was launched, Jefferson had written, "It may be asked why I insert the Mammoth [into a list of American mammals] as if it still existed? I ask in return, why I should omit it, as if it did not exist? Such is the economy of nature, that no instance can be produced of her having permitted any one race of her animals to become extinct, of her having formed any link in her great work so weak as to be broken."[14]

Thomas Jefferson was wrong. Nature's "great work" had indeed been broken, most recently and most onerously by us.

Paul Martin's "redemptive" proposal was the final presentation at this Mammoth Memorial Service—one of several events during a weekend of anniversary celebrations in Hot Springs. The proposal had been aired in a conservation journal a few months earlier, with biologist David Burney as coauthor.[15] Here in South Dakota Paul was making the first oral presentation of the idea.

"The responsibility lies on us," he told the gathering, "to restart the evolution of an extinct order, the proboscideans: the elephants, the mammoths, the mastodons. We can't have them back as we see them in the fossil record, but we have in Asia and Africa their living relatives that are in trouble in many countries on those two continents. Whether or not they pull through with international help, we still have a continent where these animals once ranged—where, with introduced proxies or surrogates, we have the opportunity to restart the evolution of this lineage.

"This is long-term stuff," he declared. "This is long beyond my lifetime and yours."

Paul Martin, of Quaker and Schwenkfelder heritage, is a gentle speaker, even when presenting content that would lend itself to fist pounding. "We finally have a chance to step beyond the large animals that we think are all we deserve—namely, bison, elk, and moose. We can include animals that our species saw in the New World before these animals became extinct. We as representatives of our species can respond to that."

Which animals does Paul Martin propose to resurrect?

"The fossils most commonly found in Pleistocene sediments are the bones of extinct horses, mammoths, mastodons, and camels. I'd love to bring back the ground sloths, too, but there are no survivors any bigger than the little tree sloths of the tropics that weigh ten, twenty pounds."

He clicked to a slide of the familiar painting by Charles Knight of Pleistocene mammals. Ground sloths, condors, and a sabertooth surround a tar pit in what is now California, with mastodons and other big mammals visible in the distance. "Here's a vision of what Rancho La Brea looked like at one time. We can't bring back the sabertooth cat at any functional level. But people who are interested in wilderness and in a wild America, and who want to get carnivores back into the landscape, have a chance to add to their shopping list the African lion. There was, once upon a time, a species of lion in the U.S., in North America, of similar size and taxonomically of close relation to the living lion in Africa.

"Now, we may have trouble as it is, coexisting with coyotes, wolves, and bears. But if we're thinking about a wild America—of dedicating a segment of the country to large animals of all kinds that have a right at some taxonomic level to be considered in this vision of a Pleistocene Park—then the lion is a species that should be on the shopping list, too."

Thirty years ago Paul Martin had urged the return of camels to American deserts.[16] His aim then was not to restart evolution in the Western Hemisphere; rather, the recommendation was remedial and practical. The West was suffering from invasion of grasslands by mesquite and juniper trees, and by shrubs such as creosote bush. He suspected that a big part of the problem was the absence of the Pleistocene browsers.

Why camels? The camel family originated in North America, as did the horse family. Surely both belong here. From the perspective of deep time, these creatures are not exotic to this continent but have been forcibly estranged from it.[17] Horses are primarily grazers, however, so only camels could be expected to prune back the woody invaders of American grasslands.

Now Martin has decided that elephants, too, deserve a place within the menagerie of resurrections. Order Proboscidea originated in Africa, so the elephant clan is not as deeply American as the camel and the horse, but proboscideans did have a continuous presence in North America for a very long time. Elephants, in fact, have far more claim on the landscape than do bison. Bison wandered across the Bering land bridge from Asia less than half a million years ago. Even so, bison are more American than we are. Humans colonized the landscape only fifteen (some believe as much as thirty) thousand years ago. In contrast, for fifteen *million* years order Proboscidea has been sending ambassadors to the New World, who then left descendants shaped by their adopted landscapes.

In a kind of parallel evolution at the cultural level, the 1990s ushered in more than one proposal for Pleistocene restoration. A Pleistocene Park of modest content but vast ambition is under way in Siberia, initiated by the Russian scientist Sergei Zimov.[18] Half a century ago, Siberia was seeded with muskox from Canada, which had been extinct in Eurasia for three thousand years. Zimov now wants to do more. He hopes to stock the park with Canadian wood bison as well. These arctic-adapted bison were hunted to extinction in Asia two thousand years ago, but they hung on in a remote patch of northern Canada.[19] Pleistocene Park is now home to Yakutian horses, and Zimov hopes to introduce reindeer and moose along with wood bison.

This park is a scientific as well as restorative venture. Glacier-free because of aridity, Siberia had been a grassy steppe rather than a mossy tundra during the Pleistocene.[20] The park will allow Zimov to test his hypothesis that when hooved mammals vanished, the arctic steppe was doomed. Mainstream theory goes the other way: invasion of grasslands by tundra mosses is presumed to have starved out the big grazers. Zimov predicts that the animals set loose in his park will trample rootless mosses into oblivion. Nutritious grasses will take their place, once again enriching the far north.

No one is suggesting that Zimov add Asian or African elephants to his wish list. Modern elephants are not equipped for an arctic lifestyle. But what about a real mammoth? Could scientists clone a living mammoth from a frozen carcass? Successful cloning of a mammoth and its resurrection in Siberia's Pleistocene Park would outshine Paul Martin's comparatively modest proposal to bring the elephant lineage back to America. Larry Agenbroad is one of the principals involved in current attempts to retrieve woolly mammoth tissue containing DNA fresh enough to clone. As to the ethical implications of such a project, Agenbroad quips, "I would rather have a cloned mammoth than another stupid sheep."[21]

Agenbroad helped create the anniversary celebration in South Dakota. He is the paleontologist in charge of the continuing excavation of Columbian mammoths at the site, which depends on Earthwatch volunteers. He is also the lead scientist excavating bones of a population of pygmy mammoths that had lived on the Channel Islands of California until thirteen thousand years ago.[22] Excavation of these little mammoths was the topic of his presentation at our Mammoth Memorial Service. To learn about mammoths great and small, and to bring this awareness to a broad public, is a here-and-now way to honor the dead.

It was Paul Martin, however, whose presentation transcended the science—in all its intrinsic excitement—to offer a profoundly unsettling yet irresistible vision. A year earlier, when Paul sent me the first draft of his "Bring Back the Elephants!" article, I told him that the idea was fun but crazy. Yet something about Bringing Back works its way into the soul. After a few weeks, I began to think of the idea as helpful for its sheer audacity. Such an outrageous proposal pushes the bounds of what one regards as radical. In its shadow, efforts to return wolves to a representative sample of their former range seems pedestrian. "Oh, no! Not the elephants!" I can hear my ranching neighbors protest. "Just give us the wolves." (As I write, zoo-bred Mexican gray wolves are being returned to the Gila Wilderness.)

Six months after I read Paul's draft essay, his provocative idea had become my vision too. With David Burney as coauthor, the revised essay was accepted for publication in *Wild Earth*. As a corresponding writer for this conservation journal, I submitted a companion piece, which placed the Bring Back idea within the scope of an inspiring restoration concept that had been launched in the previous issue: rewilding.[23] To rally the troops, I wrote, "If we ourselves do not bring elephants back and offer them a second chance for an evolving, deepening citizenship, then Order Proboscidea will never again produce American endemics; the evolution of Order Proboscidea in the New World will be over."[24]

Paul Martin concluded his talk at the anniversary celebration with a final appeal: "So here we have the idea of African animals with things to teach us about the distribution of woody legumes like honey locust and many other legumes in the tropics. Or the way that large animals worked our stream bottoms, our wetlands. We don't know what's going on if we think that bison are the largest animals that would occupy wetlands. What's going on in the wet areas along the rivers today may not be natural at all. It's as if we knew nothing about the role of fire in landscapes."

His slide presentation complete, Martin paused, then continued in a softer tone, "I've enjoyed this afternoon tremendously. I'm glad you all came. I particularly appreciate my friend Honey Locust coming up here from New Mexico, as an example of how education can reach out to all levels to change points of view. This continent is richer than we know. It has things to teach us that we need to know."

Remembrance

"I am Honey Locust, *Gleditsia triacanthos,* and I have to tell you that I'm shocked. I'm absolutely shocked that the mammoths are no longer with us. Until I received an invitation to present this eulogy, I had no idea the mammoths were gone."

Honey Locust was wearing an absurd headdress of spiraling pods dangling from a green plaster skullcap. Brown seeds pasted onto the plaster give the cap a spotted pattern, and lacy green fronds augment a crown of three-pronged thorns. A forest green shirt and a necklace of coinlike pods—a gift of honey locust's cousin, swamp locust—round out the costume, without which this science writer would not have had the courage to step out of character. Honey Locust was the third eulogist at the Mammoth Memorial Service, following Thomas Jefferson's press secretary and Georges Cuvier.

"If the mammoths truly are gone," she continued, "then why is it that I am still growing these cumbersome things?" Honey Locust flicked a hand at the dangling brown spirals.

"Do you like my pods?" She smiled coyly, framing her head with a cocked palm, and showing the audience both profiles. At the end of each sidelong turn, she gave her head a jiggle and the dry pods rattled. The kids in the front row giggled.

"For tens of millions of years, my ancestors perfected this method of seed dispersal. And now you humans tell me my pods are out of fashion? The mammoths are gone? Hmmmph! I mean, how was I supposed to know? I'm a street tree. I was planted in New York City amidst the concrete and auto exhaust. Why were my kind planted all over New York City? Because I'm tough. I'm Honey Locust. I'm *Gleditsia triacanthos.* But I'm sweet as sugar on the inside—on the inside of these pods.

"Now why am I sugary? Why did my kind develop sugary pulp over tens of millions of years? I'll tell you why: It's because my ancestors dis-

HONEY LOCUST COSTUME. At the Mammoth Memorial Service, Honey Locust delivered a eulogy, wearing a headdress of pods, seeds, and thorns and a necklace of devolved swamp locust pods.

covered that sweet pulp is the best way to entice mammoths and other big beasts to eat our pods. Every autumn, along would come a herd of mammoths. They'd come from miles around. The matriarch remembered the location of every female tree, because we were a very important food source when the green summer grasses had gone brown. The mammoths would gather round a tree, rise up on their hind legs, extend their trunks into the highest branches, and shake down the pods."

Honey Locust demonstrated the harvesting technique. She extended an arm toward the ceiling, shaking reddish-brown pods clutched in her fist while rattling those on her head. She went on to explain how mammoths ate the pods—and what came out the other end. "If I could send my children off into the world in a steaming, moist, rich heap of mammoth dung, I'd be one proud parent!"

Honey Locust told the gathering of her fear that without the mammoths her kind would follow the degenerate path of her cousin, Swamp Locust, who took the easy route. "Swamp Locust decided, 'Oh, I think I'll just kick back and grow a little seed pod here and there, and let the wind carry them or let the spring runoff take them down the river.' But what happens then? You're limited to where the wind blows. You're limited to where the river goes. And I tell you, river floodplains are highly competitive environments. I want something better for my kids. I want them to forest the hills."

Honey Locust then told the story of her thorns, how they once stood up to the tusks of mammoths. Sadly, she spoke of the degenerate thornless variety that has since taken to city life. Drawing the eulogy to a close, she sighed, "So yes, I'm sad, in fact I'm devastated that the mammoths are gone. But I am delighted to be here today. Because together—me with my pods and thorns, and you humans with your wonderful curiosity that inspires you to dig up bones and put them on display as a reminder of our great heritage on this continent—together, we can remember the mammoths in different ways. Mammoths live on in my pods and in your imaginations."

Honey Locust paused, then raised her voice, while an arm rattled a cluster of pods. "Rumble on, mighty mammoths, across the grassy plains of our imaginations!"

Honey Locust exited the stage and quickly returned sans headdress to resume her role as emcee. My first contribution in this Mammoth Memorial Service had been to read passages from the eulogy Aldo Leopold delivered fifty-two years earlier, at the dedication in Wisconsin of a monument to the passenger pigeon.[25] Paul Martin, who dreamed up the idea of staging a memorial service for the mammoths, had brought the Leopold eulogy to my attention after recruiting me to take the lead. We think Leopold's eulogy, as published in his *Sand County Almanac,* may be the first recorded memorial service for an extinct creature of any kind. This book may be the second.

"We have erected a monument to commemorate the funeral of a species," Leopold addressed the Wisconsin gathering in 1947. "It symbolizes our sorrow. We grieve because no living human will see again the onrushing phalanx of victorious birds, sweeping a path for spring across the March skies, chasing the defeated winter from all the woods and prairies of Wisconsin."

Other graceful and moving passages follow, but the one that dovetails with our own memorial service is this:

"Men still live who, in their youth, remember pigeons. Trees still live who, in their youth, were shaken by a living wind. But a decade hence only the oldest oaks will remember, and at long last only the hills will know."

Oh, how Leopold would have rejoiced to learn that even a young honey locust still remembers the mammoths!

Anachronistic fruits of North America.

EPILOGUE

"THERE IS ONE LAST PART TO THIS MEMORIAL SERVICE," the emcee announced to the small gathering. "A memorial service would not be complete without a hymn.

"Have you all picked up a memorial brochure? Good. The words to this hymn are printed inside. How many of you know the tune to the Beatles' *Let It Be*? Okay, I see a lot of heads nodding. This is a sing-along, if you wish. There are three verses, and if you don't know the tune to the whole song, you'll probably be able to join us for the chorus. The words for the chorus, by the way, come directly from the title of an essay Paul Martin wrote for *Wild Earth* magazine, where he presented the same ideas he spoke to us about today. He called that essay, "Bring Back the Elephants!"

Bring 'Em Back

When we look around North America,
There is something that it lacks.
It's the mighty elephant, bring 'em back.
O'er this continent they once did roam,
Plains and forest—all were home.
Bring back the elephants, bring 'em back!

Bring 'em back, bring 'em back.
Bring 'em back, bring 'em back.
Bring back the elephants, bring 'em back!

Though we find ourselves in a time of trouble,
The past can help us shape our dreams.

Dreaming of the future, let us dream.
What vision will help move us toward
A beautiful and gentle track?
Bring back the elephants, bring 'em back!

Chorus

Many times in twenty million years
The elephants have journeyed here.
From lands of the Old World, they did come.
A capacity to change their act
Is an e-vo-lu-tion-ary fact.
We owe them a future, bring 'em back!

Chorus

ACKNOWLEDGMENTS

THIS BOOK WOULD NOT BE, had not Paul Martin in 1992 chosen to publish his moving and lyrical essay "The Last Entire Earth" in one of my favorite journals, *Wild Earth.* Ever since, I have lived on a continent of ghosts. I interviewed Paul in 1996 about North American ghosts for my previous book; our initial rapport has since blossomed into a friendship as well as collaboration. I look to Paul not only as naturalist mentor but as mystic mentor, whose grasp of the science and half-century of poking around the Sonoran Desert allow him to speak for the ground sloths, urging us all to keep their memory alive.

Balancing Paul's energetic encouragement in this book project has been Dan Janzen's equally energetic criticism. I have not met Dan, but he graciously and immediately responded to my email queries and corrected long passages of text. Harsher comments I have not known, but the book is the better for it. I also benefited enormously from the invitation to visit the Anachronism Box in his office at the University of Pennsylvania, while he was away in Costa Rica. He had been stuffing relevant papers and clippings into that box for two decades, with the intent to eventually write a book. That intent has and will continue to take a backseat to Dan's overriding passion to preserve biodiversity in Costa Rica and around the world.

A third person has played an equally integral role in the genesis of this book: Bill Frucht, formerly my editor at Copernicus Books and now at Basic Books. This project was Bill's idea. During a lunch with science faculty at the University of Massachusetts at Amherst, he overheard me grilling a botanist across the table on the mysteries of ginkgo fruit. Bill saw a book taking shape somewhere in that conversation, and he even had the title for it. I was not convinced. Gradually, after several months of fits and starts in the library and discussions with Bill over lunch, I knew that a manuscript could indeed measure up to the evocative title. In the final stages, Bill again proved his editorial prowess as master of deletion, without treading too painfully on nonnegotiable idiosyncrasies of style.

For peer-reviewing portions of the manuscript, I thank Dan Janzen, John Dawson, Peter Del Tredici, C. E. Stevens, Jamie Gilardi, and Alfred Traverse. For scrutinizing the whole, I am deeply indebted to Paul Martin. Thanks, too, to Colin and Lauren Chapman, Richard Tedford, Carl Mehling, Bob Langsenkamp, Carol Buckley, Frank Schambach, Allen Campbell, and Carl Zimmer for conversations crucial to this book. For information and other assistance, I have also relied on Larry Agenbroad, Bill Barlow, Mary Burke, John Brewer, Becky Campbell, Eddie Clay, Nyree Conard, Richard Dawkins, Eric Dinerstein, Norm Ellstrand, Margaret Falk, Mrs. Foxie, Mauro Galetti, Danusha Goska, Ronda Green, Sharon Hermann, Larry Himes, Richard Holdaway, Henry Howe, Sarah Hunkins, Snake Jones, Betsy Koenigsberg, Siri Knutson, Cynthia Lindquist, Gabby and Ann Marek, Joe Muller, Gary Paul Nabhan, Kirk Pomper, George Rabb, Mark Steiner, Douglas Still, Jim Taulman, Tom Van Devender, Peter S. White, and Carlos Yamashita. Alice Colwell provided literary judgment as well as meticulous copyediting, and Vanessa Mobley handled publishing details with skill and grace.

My life work as a science writer would be impossible without several of the great libraries of New York City: Bobst Library at New York University, the library at the American Museum of Natural History, and the library at the New York Botanical Garden. Also crucial have been the reference and interlibrary loan services provided by always friendly staff at Miller Library of Western New Mexico University. And how, I wonder, did anyone ever write a science book before Science Citation Index went online?

Thanks, too, to the staff who manage (and the Ghosts of Collectors Past who haunt) the herbaria at the University of Arizona and the New York Botanical Garden. These repositories gave me experience with exotic fruits I could have had in no other way. Similarly, four botanical gardens provided crucial contact with the living: Arnold Arboretum, the New York Botanical Garden, the Brooklyn Botanical Garden, and the Fort Worth Botanical Garden.

I am blessed with a mate whose obsessive creativity makes him respectful of my own; thank you, Tyler. I am blessed, too, with a migratory life that affords access to both culture and wilderness, yet I am aware that this lifestyle is unachievable and unsustainable for the many. Previous generations may have lacked my own nascent skills in conjuring the Ghosts of Frugivores Past, yet other gifts of science have stripped us of our natural limits. And so we live on a continent of ghosts, hell-bent on raiding the rest of the living world in order to create even more. I acknowledge this probable future; I yearn for another.

NOTES

Chapter 1

1. Eleven thousand years has long been the date associated with megafaunal extinctions in the Americas. That date is in "radiocarbon years," however, which have recently been shown to understate the time in calendric years near the end of the Pleistocene. Eleven thousand radiocarbon years is thus now translated as thirteen thousand calendric years. Recalibration of the carbon-14 decay rate against the uranium decay rate in corals appears in E. B. Bard et al. 1990. Recalibration of the carbon-14 date against the annual cycles detected in varve deposits is reported in Irena Hajdas et al. 1995.

2. For a review of the end-Pleistocene extinctions, see Ross MacPhee 1999. See especially the chapter by Paul S. Martin and David W. Steadman. Percent losses of megafauna on the continents are also tallied in Paul S. Martin 1990, "Forty Thousand Years of Extinctions."

3. Ross MacPhee 1997.

4. Jared Diamond published a review essay in 1990, "Biological Effects of Ghosts." The phrase "ghosts of herbivores past" appears in Judith H. Myers and Dawn Bazely 1991. John A. Byers wrote of "the ghosts of predators past" in his 1997 book. All three publications cited Daniel H. Janzen and Paul S. Martin 1982.

5. Janzen and Martin 1982.

6. The ancestral form and habitat of the cultivated avocado, along with its dispersal traits, are discussed in Robert E. Cook 1982.

7. Jaguar fondness for avocado is reported in Cook 1982.

8. W. Hallwachs 1980, p. 304.

9. D. H. Janzen, email to C. Barlow, 20 February 2000.

10. *Relict* has long been a standard term in the biological sciences. It often signifies a taxon or population of plant or animal known to have been numerous and widespread in a former time but persisting today as a mere fragment of its former glory. *Webster's Collegiate Dictionary,* tenth edition, provides this definition of *relict:* "(1) widow; (2) a persistent remnant of an otherwise extinct flora or fauna or kind of organism; (3) . . . something left unchanged." I asked Janzen about his choice of terminology. He replied, "I used *anachronism* because *relict* is jargon and *anachronism* was, in my youth, a street word." He added that *anachronism* is "part of my normal vocabulary. My mother was an English teacher." *Relict* and *anachronism,* in the ways that John Byers in 1997 and Janzen and Martin in 1982 used these terms, are synonyms. They evoke the same kind of consequence stemming from the same class of evolutionary event—the latter being the extinction of animals that once exerted directional selection on the trait of an organism.

11. Dan Janzen (email to C. Barlow, 11 December 1999) strongly disagreed with the pertinence for the anachronism concept of Byers's term "overbuilt," and he criticized my portrayal of anachronisms for containing "anthropomorphisms." Paul Martin (review of manuscript), however, reacted favorably to the portrayal of anachronisms as "overbuilt," and he did not criticize my explanatory text. Botanist Peter Del Tredici offered a softer version of the overbuilt idea in a taped interview. He preferred to think of such characteristics as "overcommitted in one direction."

12. Byers 1997, p. 243.

13. Richard Dawkins 1998.

14. Dawkins 1982, p. 36.

15. R. M. Greenwood and I. A. E. Atkinson used the term *evolutionary anachronism* casually in their 1977 paper on moa and divaricate plants (p. 24), though not for the core examples in their paper.

16. D. H. Janzen, email to C. Barlow, 21 November 1999.

17. *Spandrel* appears in Stephen J. Gould and Richard C. Lewontin 1979. *Exaptation* appears in Stephen J. Gould and Elizabeth Vrba 1982. See also Gould 1997.

18. Charles Darwin, *On the Origin of Species,* first edition, chapter 6 ("Difficulties of the Theory"), section titled "Organs of Little Apparent Importance."

19. Darwin, chapter 4 ("Natural Selection"), section titled "Circumstances Favourable. . . ."

20. Darwin, chapter 5 ("Law of Variation"), section titled "Economy of Growth." In the sixth edition, the language is toned down. The phrase "any diminution, however slight, in its development will be seized on by Natural Selection" becomes simply "its diminution will be favoured" (the rest of the sentence is unchanged).

21. Martin and Burney 1999.

22. Paul Martin presents his overkill idea in his 1966 and 1967 papers in *Nature*.

23. Julia Ellen Rogers 1917, p. 180.

24. A 1983 commentary by Nils Stenseth can be read as support for research on anachronisms: "Interpretations about adaptiveness must be made with great care, and many of the discussions about optimality in evolutionary plant ecology will have to be reconsidered. However, it is also a danger to adopt the adaptationist critique in absurdum. The progress of theory in evolutionary ecology will be hindered if one abandons adaptive interpretations and falls into phenomenology. Instead of false interpretations that cause discussion and progress, we might face long fatiguing descriptions without any attempts at causal explanation."

Chapter 2

1. Janzen's warning about the extinction of ecological interactions appeared in his 1974 article "The Deflowering of Central America."

2. Janzen's role in the establishment of Guanacaste National Park is presented in William H. Allen 1988; see also Marjorie Sun 1988.

3. The presence of extinct Pleistocene megafauna in New Guinea was virtually unknown until the year after Janzen and Martin published their anachronism paper. For a summary of Australia and New Guinea Pleistocene megafauna, see T. F. Flannery and R. G. Roberts 1999.

4. Dennis J. O'Dowd and A. Malcolm Gill reported in 1986 that "evidence for seed dispersal in Australian acacia is scant, and with a few exceptions, is primarily anecdotal." In 1982 two Australian scientists (David Symon and G. D. Stocker) corresponded with Janzen about possibilities.

5. O'Dowd and Gill 1986, pp. 116–17.

6. D. W. Davidson and S. R. Morton 1984.

7. Early papers that strain for adaptive explanations include Daniel H. Janzen 1971, "Escape of *Cassia Grandis* Beans from Predators in Time and Space"; Don E. Wilson and Daniel H. Janzen 1972, "Predation on *Scheelea* Palm Seeds by Bruchid Beetles"; and Janzen 1975, "Intra- and Interhabitat Variations in *Guazuma Ulmifolia* Seed Predation."

8. The results of tests by Dan Janzen of anachronistic features of individual fruits are reported in his 1979 and 1981 papers on *Enterolobium cyclocarpum* (guanacaste), his 1982 paper on *Guazuma ulmifolia,* his 1982 paper on *Crescentia alata,* and his 1985 paper on *Spondias mombin.*

9. D. H. Janzen, email to C. Barlow, 8 December 1999.

10. Janzen, email to C. Barlow, 8 December 1999.

11. Janzen 1982, "Seeds in Tapir Dung."

12. All of the data and quotations attributed to Janzen pertaining to *Crescentia alata* are derived from Janzen and Martin 1982 or Janzen 1982, "Fruit Traits and Seed Consumption by Rodents of *Crescentia Alata,*" or Janzen 1982, "How and Why Horses Open *Crescentia Alata* Fruits."

13. Janzen 1981, "Guanacaste Tree Seed-Swallowing by Costa Rican Range Horses." Also, his 1981 "*Enterolobium Cyclocarpum* Seed Passage Rate and Survival in Horses, Costa Rican Pleistocene Seed Dispersal Agents." For a full list of citations on the follow-up research, see Claudia M. Campos and Ricardo A. Ojeda 1997.

14. Janzen's 1981 *Enterolobium* paper, p. 600.

15. Paul Martin, letter to Dan Janzen, 26 May 1978.

16. Maxine F. Miller 1996. *Acacia erioloba* is another analog; see Joseph P. Dudley 1999.

17. Nathaniel T. Wheelwright and Gordon H. Orians 1982.

18. Richard Dawkins 1982, p. 46.

19. Daniel H. Janzen 1970, "Herbivores and the Number of Tree Species in Tropical Forests"; also his 1971 "Seed Predation by Animals."

20. Anna Traveset 1990.

21. Carlos M. Herrera 1989, "Vertebrate Frugivores."

22. H. F. Lamprey, G. Halevy, and S. Makacha 1974. More recently, Christoph Rohner and David Ward 1999.

23. Janzen 1971, "Seed Predation by Animals," p. 470.

24. Janzen 1975, *Ecology of Plants in the Tropics*, p. v.

25. George Gaylord Simpson commented on the draft paper in a personal letter to Janzen, dated 7 March 1980. Simpson wrote, "Your hypothesis is summarized on p. 24 as being that the fruit fall of neotropical trees is 'intended' for a megafauna that is now extinct. That 'intended' is one of a number of, to me, objectionable anthropomorphisms in this paper." Personal files of Paul Martin.

26. In 1982 the radiocarbon dating system for Pleistocene bones had not yet been accurately calibrated to calendric years—hence, Janzen's use of ten rather than thirteen thousand years.

27. Andrew Schnabel and Jonathan F. Wendel 1998.

28. Michael A. Gold and James W. Hanover 1993.

29. Gold and Hanover 1993, p. 190.

30. Eliot Tozer 1980.

31. Andrew Schnabel and J. L. Hamrick 1990.

32. Philip J. Burton and F. A. Bazzaz 1991.

33. Gold and Hanover 1993, p. 190.

34. S. M. Csurhes and D. Kriticos 1994.

35. Charles B. Heiser Jr. 1985.

36. Gold and Hanover 1983, p. 195.

37. Tozer 1980, p. 42.
38. Gold and Hanover 1993, p. 197.
39. Gary Haynes 1991; Dudley 1999.
40. Tozer 1980, p. 52.
41. Csurhes and Kriticos 1994, p. 102.
42. Gold and Hanover 1993, p. 190.
43. Julia Ellen Rogers 1917, p. 180.
44. Rutherford Platt 1952, p. 52.
45. Sargent quoted in Heiser 1985, p. 183.
46. Heiser 1985, p. 184.
47. James C. Kroll, email to C. Barlow, 12 April 1999.
48. A. Schnabel, J. D. Nason, and J. L. Hamrick 1998.
49. Andrew Schnabel, Roger H. Laushman, and J. L. Hamrick 1991.
50. Andrew Schnabel, email to C. Barlow, 15 March 2000.
51. Pedro Jordano 1995.
52. Schnabel and Wendel 1998.

Chapter 3

1. A summary of the early stages of seed dispersal science is presented in Nathaniel T. Wheelwright and Gordon H. Orians 1982. L. Van der Pijl 1972 is the first widely cited publication on seed dispersal. In his 1971 paper "Seed Predation by Animals," Dan Janzen notes the problems in the literature caused by a widespread failure to distinguish between target seed dispersers and seed predators when mention is made of a "seed eater" (p. 471).
2. Daniel H. Janzen 1986, "Mice, Big Mammals, and Seeds."
3. Carlos M. Herrera 1985 uses the word *guild* in discussing dispersal syndromes.
4. W. Bond and P. Slingsby 1984.
5. Charles H. Janson 1983.
6. Mary F. Willson shows a preference for a lumped "vertebrate dispersal system" in her 1993 paper and 1989 coauthored paper. Pedro Jordano 1992 supports a continuum approach, with birds at one end and mammals at the other.
7. Carlos M. Herrera 1989.
8. Herrera 1986, p. 11.
9. Jordano 1992.
10. Joanna E. Lambert and Paul A. Garber 1998.
11. D. H. Janzen, email to C. Barlow, 11 December 1999.
12. Janzen 1970, "Herbivores and the Number of Tree Species," and Janzen 1971, "Seed Predation by Animals."
13. Janzen, email to C. Barlow, 11 December 1999.

14. Paul Martin's megafaunal dispersal syndrome has been applied to Africa by Diana Lieberman and Milton Lieberman (1987) and by François Feer (1995). Jose M. V. Fragoso and Jean M. Huffman (2000) portray Martin's thesis as "supported" by subsequent work in Africa.

15. Paul S. Martin, email to C. Barlow, 29 February 2000.

16. R. Norman Owen-Smith 1988, p. 1.

17. Janzen and Martin 1982 does not parse the dozen traits of the megafaunal dispersal syndrome into the three broad categories that I list here, and I have added several traits.

18. Jordano 1992 defines a "legitimate" dispersal agent as one who "swallows whole fruits and defecates or regurgitates seeds intact" (p. 122).

19. The floodplain indicator was not mentioned in Janzen and Martin 1982, but an early draft of that paper included this specification of what was then called the megafauna coevolutionary syndrome: "If planted, the trees can grow in many more local habitats than they will occupy in relatively undisturbed forest."

20. Alwyn H. Gentry 1983.

21. D.-Y. Alexandre 1978 lists as Type 1 African elephant fruits *Panda oleosa* (family Pandaceae), *Sacoglottis gabonenensis* (Humiriaceae), and various species of *Parinari* (Rosaceae). African fruits of Type 2 are represented by *Pentadesma butyraceae* and *Mammea africana* (Guttiferea), *Hirtella butayei* (Rosaceae), and three species within family Sapotaceae: *Kantou guereensis, Tieghemella hecklii*, and *Endotricha taiensis*. Exemplars of Type 3 include two species of *Tetrapleura* within the mimosa subfamily and *Swartzia fistuloides* within the caesalpinoides subfamily.

22. Takakazu Yumoto and Tamaki Maruhashi 1995; also Joseph P. Dudley 2000.

23. Lee J. T. White et al. 1993. The three most abundant species of seeds in elephant dung at this field site were *Duboscia macrocarpa* (family Tiliaceae), *Klainedoxa gabonensis* (family Irvingiaceae), and the legume pod *Swartzia fistuloides*.

24. François Feer 1995, "Morphology of Fruits."

25. Eric Dinerstein and Chris M. Wemmer 1988.

26. Richard T. Corlett 1998.

27. Corlett 1998. Rhinos are portrayed as the prime dispersal agent of mango in Malaysia in Junaidi Payne 1990. T. Volk (personal communication) collected local reports of mango sprouting from elephant dung in Thailand.

28. Lauren J. Chapman, Colin A. Chapman, and Richard Wrangham 1992.

29. White et al. 1993 and G. Klaus et al. 1998.

30. William C. Mahaney et al. 1997.

31. Dan Janzen speculated that in some legume seeds the embryo may be free of the toxins contained in the seed's endosperm. See p. 926 of his 1977 cowpea

paper. Germination and other cultural techniques for deactivating plant toxins are discussed in E. Nwokolo and J. Smartt 1996.

32. A. Gautier-Hion et al. 1985.

Chapter 4

1. The fruit of the Brazilian babassu palm, *Orbignya,* is portrayed as anachronistic in Anthony Anderson and Michael Balick 1988.

2. The early fruit-specific papers by Dan Janzen that required reinterpretation after he had his anachronism insight in 1977 include those on *Dioclea megacarpa* (1971), *Cassia grandis* (1971), *Scheelea rostrata* (1971, 1972), *Guazuma ulmifolia* (1975), *Hymenaea courbaril* (1976), *Andira inermis* (1976), and *Pithecellobium saman* (1977). The fruit-specific papers by Dan Janzen that test the anachronism idea are those on *Enterolobium* (1981, 1982), *Crescentia alata* (1982), *Pithecellobium saman* (1982), *Andira* (1982), and *Spondias mombin* (1985). A colleague of Janzen's, W. Hallwachs, reinterpreted *Hymenaea courbaril* in 1986.

3. Jörg U. Ganzhorn et al. 1999 speculate that some of the largest fruits in Madagascar may have been targeted for dispersal by a recently extinct lemur that was the size of a gorilla. John Dransfield suspects that the fruits of at least two palms *(Voianiola gerardii* and *Satranala decussilvae)* may have evolved for dispersal by Madagascar's extinct elephant birds (email to C. Barlow, 7 April 2000).

4. Quotation by Van der Hayen (1649), as reported in Patrick Armstrong 1995.

5. Berthold K. P. Horn 1978.

6. S. H. Hnatiuk 1978.

7. Charles Rick and Robert Bowman 1960.

8. Papers that mention the importance of pulp removal to prevent seed rot include: Daniel H. Janzen and Paul S. Martin 1982, Anna Traveset 1990, Mary F. Willson 1997, J. C. Noble 1975, Henry F. Howe and Gayle A. Vande Kerckhove 1981, Desmond R. Layne 1996, Jenny Ladley and Dave Kelly 1996, and Paulo Oliveira et al. 1995.

9. For more speculation by scientists working on Mauritius, see David Quammen 1996, pp. 345–53.

10. D. H. Janzen 1979, "New Horizons in the Biology of Plant Defenses," p. 339.

11. Henry F. Howe 1980.

12. Howe and Vande Kerckhove 1981.

13. Walter Hartwig and Castor Cartelle 1996.

14. D. H. Janzen 1979, "How to Be a Fig."

15. A. Gautier-Hion et al. 1985.

16. Howe 1986.

17. Howe 1986. The fruit of *Gustavia superba* was studied in detail for a report (Victoria Sork 1987) published well after the anachronism idea had been launched, but there is no indication the author considered the role extinct megafauna might have played.

18. Howe 1989.

19. Howe 1986, p. 153. The persistence problem for anachronisms was raised again in Howe 1989.

20. Howe 1985.

21. D. H. Janzen 1982, "Seeds in Tapir Dung."

22. José M. V. Fragoso and Jean M. Huffman 2000.

23. Janzen 1982, "Seeds in Tapir Dung."

24. Janzen 1982, "Wild Plant Acceptability to a Captive Costa Rican Baird's Tapir." Also, Janzen 1981, "Digestive Seed Predation by a Costa Rican Baird's Tapir."

25. Janzen 1981, "Digestive Seed Predation," p. 62.

26. Fragoso and Huffman 2000.

27. N. Smythe 1986 (p. 180) develops a similar gradational view. Janzen and Martin 1982 is cited.

28. D. H. Janzen, email to C. Barlow, 7 December 1999.

29. Janzen 1982, "Seeds in Tapir Dung," p. 132.

30. Pedro Jordano 1992 (p. 122) defines *legitimate disperser*.

31. Robert E. Cook 1982.

32. Stephen Jackson and Chengyu Weng 1999.

33. Alwyn Gentry 1983.

34. D. H. Janzen, email to C. Barlow, 7 December 1999.

35. Frank Vasek 1980.

36. The ability for clonal species to live a thousand or more years is reviewed in W. J. Bond 1994. The importance of vegetative forms of reproduction to the persistence of plant taxa through evolutionary time is explored in Alfred Traverse 1988.

37. Peter Del Tredici 1995.

38. Ludwig Kammesheidt 1999.

39. J. M. van Froenendael et al. 1996.

40. Bond 1994.

41. Mary Willson and Douglas Schemske 1980.

42. Eliane Norman, Kathleen Rice, and Steven Cochran 1992; also, Eliane Norman and David Clayton 1986.

43. Gary Paul Nabhan 1987, "Origins of Neotropical Horticulture"; also pp. 13–15 in Nabhan 1989.

44. Margaret Falk, email to C. Barlow, 5 May 2000.

Chapter 5

1. Paul Martin, in comments on a draft of this manuscript, wrote, "I'm told that tomato and chili pepper seeds germinate in the effluent of sewage treatment plants around Tucson."

2. The description that unripe persimmon "shrivels your mouth painfully" is drawn from Rutherford Platt 1952, p. 28. That it is "so puckery" comes from Julia Ellen Rogers 1917, p. 175. The quotation by Captain John Smith is drawn from Donald Culross Peattie 1950, p. 681.

3. The endangered tree is *Diospyros angulata,* as reported in Nigel Williams 1998.

4. James F. Taulman and James H. Williamson 1994. Taulman confirmed by email that the seeds had been swallowed and defecated by the captive raccoons.

5. Lynn Rogers and Roger Applegate 1983.

6. Peattie 1950, p. 682.

7. On box turtles, see W. D. Klimstra and Frances Newsome 1960.

8. Carlos M. Herrera 1989, "Frugivory and Seed Dispersal by Carnivorous Mammals."

9. Janzen employs the terminology "safe sites" on p. 509 of his 1970 "Herbivores and the Number of Tree Species in Tropical Forests."

10. R. O. Bustamante, J. A. Simonetti, and J. E. Mella 1992.

11. Harold C. Reynolds 1945 reports the misidentification of raccoon dung as possum dung. Accordingly, I wonder if the same misidentification of "opossum" dung containing persimmon seeds was made in C. Brooke Worth 1975. Carl G. Hartman reported in 1952 that "the opossum, unlike the raccoon, rarely swallows the persimmon pit" (p. 62).

12. D. H. Janzen, email to C. Barlow, 8 December 1999. Paul Martin reported in comments on the draft manuscript that "a rare species of *Diospyros* in and around Alamos, Sonora, is found only in localities where it appears to have been planted by humans"; he cited H. S. Gentry 1942.

13. I witnessed young persimmons growing abundantly along fences in a private game reserve in South Carolina; the reserve manager told me that raccoons had probably dispersed the seed. I have also seen more than a few fruit-bearing trees in a seasonally inundated floodplain in southern Indiana.

14. D. H. Janzen, email to C. Barlow, 8 December 1999.

15. Diana Lieberman and Milton Lieberman 1987.

16. Caroline Tutin et al. 1996.

17. Pedro Jordano 1992, p. 122.

18. William Clark, entry of 18 September 1806 in *The Journal of the Lewis and Clark Expedition.*

19. Dominique Caparros-Lefebvre et al. 1999.

20. Desmond Layne 1996.

21. Mary Willson and Douglas Schemske 1980. Pollination difficulties in related *Asimina* species are reported in Eliane Norman, Kathleen Rice, and Steven Cochran 1992 and Eliane Norman and David Clayton 1986.

22. J. R. Kohn and B. B. Casper 1992.

23. On the seeds in particular, see E. Nwokolo 1996, "Bottle Gourd."

24. Gary Paul Nabhan 1987, *Gathering the Desert,* p. 168.

25. The medicinal value of desert gourds to indigenous North Americans is presented in Nabhan's 1987 *Gathering the Desert,* pp. 166–82.

26. Lee Newson, S. David Webb, and James Dunbar 1993.

27. Takakazu Yumoto and Tamaki Maruhashi 1995 and Joseph P. Dudley 2000.

28. Nabhan 1987, *Gathering the Desert,* p. 177.

29. Eric Dinerstein and Chris Wemmer 1988.

30. Seeds of *Cucurmis manii* were observed in elephant dung by Martin Tchamba and Prosper Seme 1993. Colin and Lauren Chapman also report cucumber relatives in the diet of forest elephants (personal communication).

31. Nabhan 1987, *Gathering the Desert,* pp. 166–82.

32. M. A. Osman and C. W. Weber 1995.

33. Dan Janzen wrote that A. H. Gentry in 1974 "overlooked the role of Pleistocene megafauna to such a degree that he viewed *Crescentia* as water-dispersed despite its sweet fleshy interior" (in Janzen 1982, "Fruit Traits and Seed Consumption by Rodents of *Crescentia Alata*"). Similarly, J. Robert Hunter (1989) criticized Janzen and Martin 1982 by concluding that the dispersal strategy of *Enterolobium cyclocarpum* is flotation.

34. Janzen 1982, "Fruit Traits and Seed Consumption by Rodents of *Crescentia Alata.*"

35. The concept was later published in brief in Nabhan's 1989 *Enduring Seeds,* pp. 13–15.

36. Kohn and Casper 1992.

37. W. Hallwachs 1986. See also Nigel M. Asquith et al. 1999.

Chapter 6

1. Donald Leopold et al. 1998, p. 242. Also, G. H. Collinwood 1939.

2. Jeffrey Smith and Janice Perino 1981.

3. Smith and Perino 1981.

4. Perhaps squirrels long familiar with the fruit are less repelled by the latex. Andrew Schnabel (email to C. Barlow, 15 March 2000) reports that "it is not difficult to find fresh maclura fruits that have been completely picked apart by squirrels for the seeds. The sticky latex apparently does not bother them in the least."

5. Joseph P. Dudley 1999.

6. Meriwether Lewis, letter to President Thomas Jefferson, 4 May 1804.

7. Donald Peattie 1950, p. 479. The theory that osage orange had retrenched to just Bois d'Arc Creek is proposed in Frank F. Schambach 2000.

8. William Werthner 1935, p. 189.

9. Smith and Perino 1981.

10. Andrew Schnabel, Roger H. Laushman, and J. L. Hamrick 1991.

11. Peter Del Tredici told me that maclura can produce fruits and seeds apomictically.

12. More than three hundred plant species, in at least thirty taxonomic families, can reproduce by apomictic clones (Karl Niklas 1997, p. 69).

13. Smith and Perino (1981) depict *Maclura* and *Cudrania* as subject to hybridization; Peter Del Tredici (personal communication) disputes this conclusion, based on his experience with the two genera at Arnold Arboretum and upon the lack of a herbarium sheet to document such a cross. Del Tredici, in turn, references Frank Santamour of the U.S. National Arboretum in Washington, D.C.

14. John Brewer 1992, p. 30.

15. Richard Corlett 1998.

16. E. Nwokolo 1996, "African Breadfruit."

17. François Feer 1995, "Morphology of Fruits"; also, Lee White et al. 1993.

18. Frank F. Schambach 2000.

19. William Werthner 1935, p. 256.

20. Werthner 1935, p. 259.

21. Nancy Turner and Adam Sczawinski 1991, pp. 79–80.

22. H. A. Stephens 1980, p. 52.

23. T. Konoshima et al. 1995.

24. D. H. Janzen 1976, "Effect of Defoliation on Fruit-Bearing Branches of the Kentucky Coffee Tree." Roasting destroys some toxins, but protein-based toxins (such as hemagglutinins, proteinase inhibitors, and impostor amino acids) usually require boiling water to allow heat-induced hydration, and alkaloids and tannins can be deactivated only by adsorption onto clay surfaces or onto one another.

25. Colin Chapman, personal communication, 30 December 1999.

26. Werthner 1935, p. 259.

27. S. Rehr, E. Bell, D. Janzen, and P. Feeny 1973. In email to C. Barlow, 26 January 2000, Dan Janzen wrote, "I suspect that if *Gymnocladus* grew in Costa Rican dry forest, it would have a weevil (not a bruchid) [as a predator], but if it had a bruchid, it probably would suffer too heavy seed predation to survive (given the small crop size)."

28. Peter Del Tredici, Hsieh Ling, and Guang Yang 1992.

29. Del Tredici et al. 1992.

30. Del Tredici 1991.

31. Chapters on ginkgo's resistance to insect and fungal attack can be found in T. Hori et al. 1997.

32. Peter Del Tredici 1993.

33. Del Tredici 1992 and 1997.

34. Del Tredici 1991.

35. J. Sastre et al. 1998; also, W. Juretzek 1997.

36. Peter Del Tredici 2000.

37. Richard Corlett 1998.

38. Nathaniel Wheelwright and Gordon Orians 1982.

39. Peter Del Tredici told me that reports from China indicate that small carnivores will consume ginkgo fruits even after they have become rank.

40. Del Tredici 1991; also, Keiji Wada and Masanobu Haga 1997.

41. Del Tredici 1993.

Chapter 7

1. William Matheny 1931, p. 76.

2. Paul Martin, letter to Dan Janzen, 15 May 1984.

3. Gary Nabhan 1987, pp. 136–48.

4. John Brewer, email to C. Barlow, 17 November 1999.

5. Janzen discusses the anachronistic features of jumping cholla in his 1986 "Chihuahuan Desert Nopaleras."

6. Janzen 1986, "Chihuahuan."

7. B. B. Simpson 1972, p. 202.

8. J. R. Brown and Steve Archer 1989; also, Steve Archer 1989.

9. Paul Martin 1975; also, J. R. Brown and Steve Archer 1987.

10. C. E. Fisher 1972, p. 178.

11. Fisher 1972, p. 178.

12. R. Norman Owen-Smith 1987. The role of elephants in tree reduction is discussed in several chapters in A. Sinclair and Peter Arcese 1995.

13. D. H. Janzen, email to C. Barlow, 8 December 1999.

14. Paul Martin 1970.

15. Martin 1990, "Thinking Like a Canyon."

16. Martin 1970.

17. Martin 1969.

18. Gerald Carson 1980; also, Jean Seligmann and Bill Hart 1988.

19. Martin (1969) quoted Beale's experience with camels and creosote bush. See also Carson 1980.

20. David Western 1997.

21. D. H. Janzen 1982, "Fruits for Famished Mammoths."

22. R. F. Mueller 1990.

23. Peter White 1988.

24. Peter White, phone conversation with C. Barlow, May 1999.

25. D. H. Janzen 1982, "Fruits for Famished Mammoths," p. 182.

26. Turner is quoted in Jeffrey Smith and Janice Perino 1981.

27. William Least Heat-Moon 1991, p. 281.

28. Alan Cooper et al. 1993.

29. M. S. McGlone and C. J. Webb 1981.

30. R. M. Greenwood and I. A. E. Atkinson 1977.

31. Helpful background on the history of the divaricate plant debate can be found in M. S. McGlone and B. D. Clarkson 1993 and in John Dawson 1993.

32. McGlone and Webb 1981.

33. McGlone and Webb 1981.

34. Dawson 1993, p. 139.

35. Strong support for rapid overkill of moa is presented in R. N. Holdaway and C. Jacomb 2000. See also the review article by Jared Diamond (2000).

36. C. J. Burrows 1980.

37. Jamie Day and Kevin Gould 1997.

38. Dawson 1993, p. 144.

39. McGlone and Clarkson 1993.

40. Cooper et al. 1993. Dawson (1993, p. 146) also judged "multiple factors" to have been at work. A paper that purports to refute the climate hypothesis is D. Kelly and M. Ogle 1990.

41. I. A. E. Atkinson and R. M. Greenwood 1989.

42. David Quammen 1998.

Chapter 8

1. My conversation with Richard Tedford took place eight months before Frank Schambach told me that horses "relish" osage orange.

2. François Feer 1995 reports seed spitting in ruminant duikers of Gabon.

3. Carlos Yamashita 1997.

4. A large literature has accrued on the fates of seeds dispersed in rumen ejecta and animal dung. The papers present conflicting conclusions as to whether burial in dung encourages or discourages seed predators (especially mice) and whether the dung itself is harmful to the tissues of newly sprouted seedlings. These papers include: D. H. Janzen 1982, "Removal of Seeds from Horse Dung by Tropical Rodents"; Janzen 1982, "Attraction of *Liomys* Mice to Horse Dung"; Anna Traveset 1990; Joseph P. Dudley 1999.

5. D. H. Janzen 1982, "Differential Seed Survival and Passage Rates in Cows and Horses."

6. Janzen 1981, "Guanacaste Tree Seed-Swallowing by Costa Rican Range Horses."

7. Ruminants have a narrow passage between the reticulorumen and omasal segments of the forestomach. The constriction in camels is where the forestomach joins the secretory stomach that produces digestive enzymes. Among hindguts the sphincters or valves that separate stomach from intestine would be the limiting factor. (Letter from C. E. Stevens to C. Barlow, 23 June 2000.)

8. R. Dale Guthrie 1990, p. 257.

9. Maxine Miller 1993.

10. D. H. Janzen 1985, "*Spondias Mombin* Is Culturally Deprived in Megafauna-Free Forest," and José M. V. Fragoso and Jean M. Huffman 2000.

11. R. T. Wilson 1989.

12. Anna Traveset 1990.

13. Dan Janzen judges cattle to be better dispersal agents than horses for guanacaste seeds in his 1985 paper "How Fast and Why Do Germinating Guanacaste Seeds Die Inside Cows and Horses?" In contrast, Claudia M. Campos and Ricardo A. Ojeda (1997) determined horses to be the better dispersal agents of mesquite seeds. For these sizes of seeds, however, differences between horses and cattle are probably insignificant; rather, both are effective dispersal agents. Dan Janzen's empirical findings are somewhat at odds with what one might construe from anatomy alone: the orifice separating forestomach chambers in cattle is estimated to pass objects with diameters no greater than half a centimeter (Burk A. Dehority 1997).

14. Richard T. Corlett 1998.

15. F. Kruelen 1985.

16. Christine Janis 1976.

17. Several adaptive hypotheses for the evolution of the ruminant digestive system are discussed in Richard E. Bodmer 1989.

18. For much of the background on forestomach versus hindgut digestive systems, I have drawn upon C. Edward Stevens and Ian D. Hume 1995.

19. W. J. Freeland and D. H. Janzen 1974.

20. Mary Burke, manager of mammals at the Brookfield Zoo, email to Paul Martin, 9 January 2000.

21. Timothy Johns 1990, p. 40.

22. The website of the Elephant Sanctuary is www.elephants.com.

23. Staff at the Brookfield Zoo in Chicago informed me later that their African elephants also consume clay "when they have yard access." Mary Burke, email to C. Barlow, 9 March 2000.

24. Gregor Klaus et al. 1998.

25. Various mammals are documented as visiting clay (or "mineral") licks in R. Krishnamani and W. C. Mahaney 2000, Mahaney 1999, Mahaney et al. 1995, Mahaney et al. 1997, Klaus et al. 1998, R. G. Guggiero and J. M. Fay 1994, and F. Kruelen 1985.

26. Jared Diamond et al. 1999 cites many papers that link different animals to geophagy. James Gilardi et al. 1999 provides empirical evidence that ingestion of clay reduces the amount of plant alkaloids circulating in the blood of parrots.

27. D. H. Janzen 1977, "Why Fruits Rot, Seeds Mold, and Meat Spoils."

28. Janzen 1982, "Wild Plant Acceptability to a Captive Costa Rican Baird's Tapir."

29. Susan Allport 2000, p. 110.

30. E. Nwokolo and J. Smartt 1996, p. 9.

31. Johns 1990.

32. Humans can roast seeds at high heats to deactivate some toxic compounds; we can boil in water (especially in a pressure cooker) to deactivate protein-based compounds that require hydrated heat (such as hemagglutinins, impostor amino acids, and proteinase inhibitors); and we can cook with clay to bind the toxins not affected by heat, notably the alkaloids and perhaps tannins.

33. Although I cannot corroborate the arms race and clay hypothesis, it would answer Jared Diamond's puzzlement about why a fruit pigeon (which does not grind seeds) was a daily visitor at a geophagy site in New Guinea (Diamond et al. 1999).

34. Upon questioning, Dan Janzen confirmed that the captive tapir he used in his tapir-plant experiments did not have access to a clay source (email to C. Barlow, 28 January 2000).

35. Mahaney et al. 1995.

36. Termite and leaf-cutter ant nests (the former often attached to tree trunks) are also attractive geophagy sites for New World herbivorous monkeys. See Kosei Izawa 1993; also, Edward Heymann and Gerald Hartmann 1991.

37. M. Profet 1992.

38. Mahaney et al. 2000.

39. R. K. Grigsby et al. 1999.

40. Wiley and Katz 1998.

41. Wiley and Katz 1998.

42. Jared Diamond (1999) wrote, "Do curious dirt-licking babies deserve our encouragement for their experiments with self-medication?"

43. Kruelen 1985.

44. S. Aufreiter et al. 1997; also, Johns 1990.

Chapter 9

1. Paul Martin and David Steadman 1999, p. 49.

2. Daniel Janzen 1988.

3. Stephen Buchmann and Gary Paul Nabhan 1996. For the vulnerability of South African plants to lost pollinators, see W. J. Bond 1994.

4. The phrase "extinction of the massive" appears in Paul Martin and David Burney 1999.

5. Stephen Jackson and Chengyu Weng 1999.

6. The genus of shrub discovered in Arizona is *Apacheria,* reported in C. T. Mason 1975 (information supplied by Paul Martin).

7. Exceptions include discovery of *Metasequoia* in China (1941) and most recently of *Wollemia* in Australia (1994).

8. The first sentence of the "Discussion" section of the original draft of their 1982 anachronism paper shows Janzen's view of the plant extinction issue in type and Martin's disagreement in pencil edits. The disagreement continues on p. 23 of Janzen's email to Barlow, 8 December 1999, and in Martin's email to Barlow, 21 February 2000.

9. Paul Martin, email to C. Barlow, 15 March 2000.

10. Dan Janzen, in his 1982 "Natural History of Guacimo Fruits," identifies two species of bruchid beetle that are parasitic on the seeds of a single species of plant *(Guazuma ulmifolia)* in Costa Rica.

11. Scott E. Miller et al. 1981.

12. Daniel Janzen 1983, "Seasonal Change in Abundance of Large Nocturnal Dung Beetles."

13. A summary of the Australian dung beetle project is reported in B. M. Doube et al. 1991.

14. Ikka Hanski 1991.

15. Steven Emslie 1987.

16. Frank Graham 2000.

17. More than a half dozen genera of scavenging birds vanished from North America at the end of the Pleistocene. For a discussion, see David Steadman and Paul Martin 1984.

18. Paul Martin 1992.

19. Daniel Janzen 1988, "Complexity Is in the Eye of the Beholder."

20. Henry Howe 1984.

21. José M. V. Fragoso and Jean M. Huffman 2000.

22. Kent H. Redford 1992. The vulnerability of tropical trees to loss of dispersers is presented in José Cardosa da Silva and Marcelo Tabarelli 2000. See also Nigel M. Asquith 1999.

23. Henry Howe 1984.

24. G. C. Stocker and A. K. Irvine 1983.

25. Kenneth Whitney et al. 1998.

26. N. Hallé 1987.

27. Caroline Tutin et al. 1991.

28. Richard Corlett 1998.

29. Jörg Ganzhorn et al. 1999.

30. Bryan Patterson 1975.

31. A. H. Whitaker 1987.

32. Because fossil bones of the Pacific rat date back two thousand years, yet evidence of human colonization appears more than a millennium later, it is presumed that Polynesians may have visited New Zealand before humans became permanent and populous dwellers on the islands (R. N. Holdaway 1996).

33. Louise Emmons et al. 1991.

34. Joseph P. Dudley 2000.

35. *Solanum elaegnifolium* is a tomato because the (South American) garden tomato has recently been renamed *Solanum lycopersicon*.

36. The fruit of *Solanum carolinense* is reported as "often persisting throughout the winter," just as I observed in the sister species, *S. elaegnifolium*. The quotation is drawn from H. A. Stephens 1980, p. 95.

37. Martin Cipollini and Douglas Levey 1997.

38. Carl H. Ernst et al. 1994, p. 274.

39. R. W. Rust and R. R. Roth 1981.

40. Joanne Braun and Garnett Brooks 1986.

41. Charles Rick and Robert Bowman 1960.

42. Tony Swain 1976.

43. J. Alan Holman 1985, p. 206.

44. Kraig Adler 1969.

45. Paul Martin and Christine Szuter 1999.

46. Paul Martin 1990, "Forty Thousand Years of Extinctions."

47. Avocado seeds are reported in the dung of African forest elephants in Martin Tchamba and Prosper Seme 1993.

48. Daniel Janzen 1985, "On Ecological Fitting."

49. S. J. Milton et al. 1990.

50. Colin Chapman and Lauren Chapman 1996.

51. Some populations of *Solanum elaegnifolium* include prickle-free variants. W. B. McDougall and Omer Sperry 1951, p. 154.

52. Daniel Janzen 1985, "*Spondias Mombin* Is Culturally Deprived in Megafauna-Free Forest."

53. Janzen 1986, "Chihuahuan Desert Nopaleras."

54. Maria Mandujano et al. 1996.

55. Andrew Schnabel (email to C. Barlow, 15 March 2000) confirms my interpretation and presentation of his views on the independent, parallel evolution of two species of *Gleditsia* in his 1998 paper with J. F. Wendel.

56. An expert on genus *Gleditsia,* Andrew Schnabel wrote in an email to C. Barlow, 15 March 2000: "I wouldn't put much stake in a hypothesis that *G. aquatica* is post-Pleistocene in origin. That would be a very rapid evolutionary change."

57. Although Andrew Schnabel and Jonathan Wendel (1998) did not consider the cause of devolution, they conclude that *G. triacanthos* is indeed the ancestor of *G. aquatica*.

58. Eliot Tozer 1980.

59. Charles Heiser 1985, p. 184.

60. Heiser 1985, p. 187.

61. Heiser 1985, p. 186.

62. S. M. Csurhes and D. Kriticos 1994.

63. Heiser 1985, p. 182.

64. Malcolm Coe and Christopher Coe 1987.

65. Schnabel and Wendel 1998.

66. David Michener 1986.

67. Csurhes and Kriticos 1994.

Chapter 10

1. Thomas Bonnicksen 2000.

2. Gary Paul Nabhan 1987, *Gathering the Desert*.

3. Daniel Janzen 1982, "Fruit Traits and Seed Consumption by Rodents of *Crescentia Alata*."

4. The conference address delivered by Gary Paul Nabhan in 1987, "The Origins of Neotropical Horticulture Following Megafaunal Extinctions: Did Humans Disperse and Select Anachronistic Fruits?" was accepted for publication, but Nabhan never followed through for completion (Nabhan, email to C. Barlow, 8 March 2000). A summary of its arguments can be found in Nabhan's 1989 book, pp. 13–16.

5. Nabhan 1987, *Gathering the Desert*.

6. Paul Martin, email to C. Barlow, 19 August 1999.

7. Thomas Berry 2000.

8. Peter Jackson et al. 1988.

9. Thanks to Peter Del Tredici, Peter S. White, and Sharon Hermann for helpful discussions on the plight of the Florida torreya.

10. Sharon Hermann at Tall Timbers Research Station in Tallahassee, Florida, provided in a phone conversation most of the information I present on the nature of torreya's problems and lack of animal dispersers.

11. Siri Knutson, office administrator for the Mammoth Site, played the ghost of Georges Cuvier.

12. The paper on extinction that Georges Cuvier read to the National Institute in 1796 ("Espèces d'éléphants") can be found in Martin Rudwick 1997.

13. The role of Thomas Jefferson's press secretary was played by Eddie Clay, a former state legislator and now president of the private association that owns and runs the Mammoth Site.

14. Thomas Jefferson 1787/1955, pp. 53–54.

15. Burney is the author of the 1993 "Recent Animal Extinctions: Recipes for Disaster."

16. Paul Martin 1969 and 1970.

17. Martin 1969.

18. Sergei Zimov et al. 1995.

19. Richard Stone 1998.

20. R. Dale Guthrie 1990.

21. Larry Agenbroad is quoted in "Going Woolly Mammoth Huntin'" 1999.

22. Larry Agenbroad et al. 1999.

23. The rewilding concept was launched in Michael Soulé and Reed Noss 1998.

24. Connie Barlow 1999.

25. Aldo Leopold's "On a Monument to a Pigeon" appears in his 1949 *A Sand County Almanac.*

REFERENCES

Adler, Kraig, 1969, "The Influence of Prehistoric Man on the Distribution of the Box Turtle," *Annals of Carnegie Museum* 41: 263–80.

Agenbroad, Larry D., et al., 1999, "Pygmy Mammoths from Channel Islands National Park," *Deinsea* 6: 89–102.

Alexandre, D.-Y., 1978, "Le Rôle disséminateur des éléphants en Forêt de Tai, Côte-d'Ivoire," *La Terre et la vie* 32: 47–71.

Allen, William H., 1988, "Biocultural Restoration of a Tropical Forest," *BioScience* 38(3): 156–61.

Allport, Susan, 2000, *The Primal Feast: Food, Sex, Foraging, and Love* (New York: Harmony Books).

Anderson, Anthony B., and Michael J. Balick, 1988, "Taxonomy of the Babassu Complex (*Orbignya* Spp.)," *Systematic Botany* 13: 32–50.

Archer, Steve, 1989, "Have Southern Texas Savannas Been Converted to Woodlands in Recent History?" *American Naturalist* 134: 545–61.

Armstrong, Patrick, 1995, "The Dodo and the Tree," *Geographical Magazine* 57: 541–43.

Asquith, Nigel M., et al., 1999, "The Fruits the Agouti Ate: *Hymenaea Courbaril* Seed Fate When Its Disperser Is Absent," *Journal of Tropical Ecology* 15: 229–35.

Atkinson, I. A. E., and R. M. Greenwood, 1989, "Relationships Between Moas and Plants," *New Zealand Journal of Ecology* 12: 76–96.

Aufreiter, S., et al., 1997, "Geochemistry and Mineralogy of Soils Eaten by Humans," *International Journal of Food Sciences and Nutrition* 48: 293–305.

Bard, E. B., et al., 1990, "Calibration of the ^{14}C Timescale over the Past 30,000 Years Using Mass Spectrometric U–Th Ages from Barbados Corals," *Nature* 345: 405–10.

Barlow, Connie, 1997, *Green Space, Green Time: The Way of Science* (New York: Springer-Verlag).

______, 1999, "Rewilding for Evolution," *Wild Earth* 9(1): 53–56.

Berry, Thomas, 2000, *The Great Work* (New York: Bell Tower).

Bodmer, Richard E., 1989, "Frugivory in Amazonian Artiodactyla: Evidence for the Evolution of the Ruminant Stomach," *Journal of the Zoological Society of London* 219: 457–67.

Bond, W. J., 1994, "Do Mutualisms Matter?" *Philosophical Transactions of the Royal Society of London*, series B, 344: 83–90.

Bond, W. J., and P. Slingsby, 1984, "Collapse of an Ant-Plant Mutualism," *Ecology* 65: 1031–37.

Bonnicksen, Thomas, 2000, *Ancient Forests: From the Ice Age to the Age of Discovery* (New York: Wiley).

Braun, Joanne, and Garnett R. Brooks Jr., 1986, "Box Turtles as Potential Agents for Seed Dispersal," *American Midland Naturalist* 117: 312–18.

Brewer, John, 1992, "The Mulberry Connection," *Hedge Apple and Devil's Claw*, spring.

Brown, J. R., and Steve Archer, 1987, "Woody Plant Seed Dispersal and Gap Formation in a North American Subtropical Savanna Woodland: The Role of Domestic Herbivores," *Vegetatio* 73: 73–80.

______, 1989, "Woody Plant Invasion of Grasslands," *Oecologia* 80: 19–26.

Buchmann, Stephen L., and Gary Paul Nabhan, 1996, *The Forgotten Pollinators* (Washington, D.C.: Island Press).

Burney, David A., 1993, "Recent Animal Extinctions: Recipes for Disaster," *American Scientist* 81: 533–41.

Burrows, C. J., 1980, "Some Empirical Information Concerning the Diet of Moas," *New Zealand Journal of Ecology* 3: 125–30.

Burton, Philip J., and F. A. Bazzaz, 1991, "Tree Seedling Emergence on Interactive Temperature and Moisture Gradients and in Patches of Old-Field Vegetation," *American Journal of Botany* 78: 131–49.

Bustamante, R. O., J. A. Simonetti, and J. E. Mella, 1992, "Are Foxes Legitimate and Efficient Seed Dispersers?" *Acta Oecologia* 13(2): 203–8.

Byers, John, 1997, *American Pronghorn: Social Adaptations and the Ghosts of Predators Past* (Chicago: University of Chicago Press).

Campos, Claudia M., and Ricardo A. Ojeda, 1997, "Dispersal and Germination of *Prosopsis flexuosa* Seeds by Desert Mammals in Argentina," *Journal of Arid Environments* 35: 707–14.

Caparros-Lefebvre, Dominique, et al., 1999, "Possible Relation of Atypical Parkinsonism in the French West Indies with Consumption of Tropical Plants," *Lancet* 354: 281–86.

Cardosa da Silva, José Maria, and Marcelo Tabarelli, 2000, "Tree Species Impoverishment and the Future Flora of the Atlantic Forest of Northeast Brazil," *Nature* 404: 72–73.

Carson, Gerald, 1980, "Jefferson Davis's Camel Corps," *Natural History* 89(5): 70–75.

Channell, Rob, and Mark V. Lomolino, 2000, "Dynamic Biogeography and Conservation of Endangered Species," *Nature* 403: 84–86.

Chapman, Colin A., and Lauren J. Chapman, 1996, "Frugivory and the Fate of Dispersed and Non-Dispersed Seeds of Six African Tree Species," *Journal of Tropical Ecology* 12: 491–504.

Chapman, Lauren J., Colin A. Chapman, and Richard Wrangham, 1992, "*Balanites Wilsoniana:* Elephant Dependent Dispersal?" *Journal of Tropical Ecology* 8: 274–83.

Cipollini, Martin L., and Douglas J. Levey, 1997, "Secondary Metabolites of Fleshy Vertebrate-Dispersed Fruits: Adaptive Hypotheses and Implications for Seed Dispersal," *American Naturalist* 150: 347–72.

Clemens, E. T., and G. M. O. Maloiy, 1982, "The Digestive Physiology of Three East African Herbivores," *Journal of Zoology* (London) 198: 141–56.

Coe, Malcolm, and Christopher Coe, 1987, "Large Herbivores, Acacia Trees, and Bruchid Beetles," *South African Journal of Science* 83: 624–34.

Collinwood, G. H., 1939, "Osage-Orange," *American Forests* 45: 508–9.

Cook, Robert E., 1982, "Attractions of the Flesh," *Natural History* 91(1): 20–24.

Cooper, Alan, I. A. E. Atkinson, William G. Lee, and T. H. Worthy, 1993, "Evolution of the Moa and Their Effect on the New Zealand Flora," *Trends in Ecology and Evolution* 8: 433–42.

Cooper, Susan M., and R. Norman Owen-Smith, 1986, "Effects of Plant Spinescence on Large Mammalian Herbivores," *Oecologia* 68: 446–55.

Corlett, Richard T., 1998, "Frugivory and Seed Dispersal by Vertebrates in the Oriental (Indomalayan) Region," *Biological Review of the Cambridge Philosophical Society* 73: 413–48.

Csurhes, S. M., and D. Kriticos, 1994, "*Gleditsia Triacanthos,* Another Thorny, Exotic Fodder Tree Gone Wild," *Plant Protection Quarterly* 9: 101–4.

Davidson, D. W., and S. R. Morton, 1984, "Dispersal Adaptations of Some *Acacia* Species in the Australian Arid Zone," *Ecology* 65: 1038–51.

Dawkins, Richard, 1982, *The Extended Phenotype* (Oxford: Oxford University Press).

______, 1998, "Science and Sensibility," *Free Inquiry,* winter, p. 39.

Dawson, John, 1993, *Forest Vines to Snow Tussocks* (Wellington, New Zealand: Victoria University Press).

Day, Jamie S., and Kevin S. Gould, 1997, "Vegetative Architecture of Divaricating Juveniles," *Annals of Botany* 79: 607–16.

Dehority, Burk A., 1997, "Foregut Fermentation," in R. I. Mackie and B. A. White, eds., *Gastrointestinal Microbiology* (New York: Chapman & Hall).

Del Tredici, Peter, 1989, "Ginkgos and Multituberculates: Evolutionary Interactions in the Tertiary," *BioSystems* 22: 327–39.

______, 1991, "Ginkgos and People: A Thousand Years of Interaction," *Arnoldia* 51(2): 2–15.

______, 1992, "Natural Regeneration of *Ginkgo Biloba* from Downward Growing Cotyledonary Buds (Basal Chichi)," *American Journal of Botany* 79: 522–30.

______, 1993, "Ginkgo Chichi in Nature, Legend, and Cultivation," *International Bonsai* 23(4): 20–25.

______, 1995, "Shoots from Roots: A Horticultural Review," *Arnoldia* 55(3): 11–19.

______, 1997, "Lignotuber Development in *Ginkgo Biloba,*" in T. Hori et al., eds., *Ginkgo Biloba: A Global Treasure* (New York: Springer-Verlag), pp. 119–26.

______, 2000, "The Evolution, Ecology, and Cultivation of *Ginkgo Biloba*," in T. van Beek, ed., *Ginkgo Biloba* (Amsterdam: Harwood Academic Publishers), pp. 7–23.

Del Tredici, Peter, Hsieh Ling, and Guang Yang, 1992, "The Ginkgos of Tian Mu Shan," *Conservation Biology* 6(2): 202–9.

Diamond, Jared, 1990, "Biological Effects of Ghosts," *Nature* 345: 769–70.

______, 1999, "Dirty Eating for Healthy Living," *Nature* 400: 120–21.

______, 2000, "Blitzkrieg Against the Moas," *Science* 287: 2170–71.

Diamond, Jared, K. David Bishop, and James D. Gilardi, 1999, "Geophagy in New Guinea Birds," *Ibis* 141: 181–93.

Dinerstein, Eric, and Chris M. Wemmer, 1988, "Fruits *Rhinoceros* Eat: Dispersal of *Trewia Nudiflora* in Lowland Nepal," *Ecology* 69: 1768–74.

Doube, B. M., et al., 1991, "Native and Introduced Dung Beetles in Australia," in Ikka Hanski and Yves Cambefort, eds., *Dung Beetle Ecology* (Princeton: Princeton University Press), pp. 255–78.

Dudley, Joseph P., 1999, "Seed Dispersal of *Acacia Erioloba* by African Bush Elephants," *African Journal of Ecology* 37: 375–85.

______, 2000, "Seed Dispersal by Elephants in Semiarid Woodland Habitats of Zimbabwe," *Biotropica* 32(3).

Emmons, Louise H., et al., 1991, "The Fruit and Consumers of *Rafflesia Keithii,*" *Biotropica* 23(2): 197–99.

Emslie, Steven D., 1987, "Age and Diet of Fossil California Condors in Grand Canyon, Arizona," *Science* 237: 768–70.

Ernst, Carl H., et al., 1994, *Turtles of the United States and Canada* (Washington, D.C.: Smithsonian Institution Press).

Feer, François, 1995, "Morphology of Fruits Dispersed by African Forest Elephants," *African Journal of Ecology* 33: 279–84.

______, 1995, "Seed Dispersal in African Forest Ruminants," *Journal of Tropical Ecology* 11: 683–89.

Fisher, C. E., 1972, "Mesquite and Modern Man in Southwestern North America," in B. B. Simpson, ed., *Mesquite* (Washington, D.C.: Institute of Ecology).

Flannery, T. F., and R. G. Roberts, 1999, "Late Quaternary Extinctions in Australasia," in Ross D. E. MacPhee, ed., *Extinctions in Near Time* (New York: Kluwer), pp. 239–55.

Fragoso, José M. V., and Jean M. Huffman, 2000, "Seed Dispersal Patterns by the Last Neotropical Megafaunal Element in Amazonia, the Tapir," *Journal of Tropical Ecology* 16: 369–85.

Freeland, W. J., and D. H. Janzen, 1974, "Strategies in Herbivory by Mammals: The Role of Plant Secondary Compounds," *American Naturalist* 108: 269–89.

Ganzhorn, Jörg U., et al., 1999, "Lemurs and the Regeneration of Dry Deciduous Forest in Madagascar," *Conservation Biology* 13: 794–804.

Gautier-Hion, A., et al., 1985, "Fruit Characters as a Basis of Fruit Choice and Seed Dispersal in a Tropical Forest Vertebrate Community," *Oecologia* 65: 324–37.

Gentry, Alwyn H., 1983, "Dispersal and Distribution of Bignoniaceae," *Sonderbd. Naturwiss. Ver. Hamburg* 7: 187–99.

Gentry, H. S., 1942, *Rio Mayo Plants* (Washington, D.C.: Carnegie Institution).

Gilardi, James D., et al., 1999, "Biochemical Functions of Geophagy in Parrots: Detoxification of Dietary Toxins," *Journal of Chemical Ecology* 25: 897–922.

"Going Woolly Mammoth Huntin'," 1999, *Science* 285: 1007.

Gold, Michael A., and James W. Hanover, 1993, "Honeylocust *(Gleditsia Triacanthos)*, a Multipurpose Tree for the Temperate Zone," *International Tree Crops Journal* 7: 189–207.

Goodall, Jane, 1998, "Learning from the Chimpanzees," *Science* 282: 2182.

Gould, Stephen J., 1997, "The Exaptive Excellence of Spandrels as a Term and Prototype," *Proceedings of the National Academy of Sciences USA* 94: 10750–55.

Gould, Stephen J., and Richard C. Lewontin, 1979, "The Spandrels of San Marco and the Panglossian Paradigm," *Proceedings of the Royal Society* B205: 581–98.

Gould, Stephen J., and Elizabeth Vrba, 1982, "Exaptation: A Missing Term in the Science of Form," *Paleobiology* 8: 4–15.

Graham, Frank, Jr., 2000, "Day of the Condor," *Audubon,* January, pp. 46–53.

Greenwood, R. M., and I. A. E. Atkinson, 1977, "Evolution of Divaricating Plants in New Zealand in Relation to Moa Browsing," *Proceedings of the New Zealand Ecological Society* 24: 21–29.

Grigsby, R. K., et al., 1999, "Chalk Eating in Middle Georgia," *Southern Medical Journal* 92: 190–92.

Guggiero, R. G., and J. M. Fay, 1994, "Utilization of Termitarium Soils by Elephants and Its Ecological Implications," *African Journal of Ecology* 32: 222–32.

Guthrie, R. Dale, 1990, *Frozen Fauna of the Mammoth Steppe* (Chicago: University of Chicago Press).

Hajdas, Irena, et al., 1995, "Problems in the Extension of the Radiocarbon Calibration Curve (10–13 KYR BP)," *Radiocarbon* 37: 75–79.

Hallé, N., 1987, "*Cola lizae* N. Hallé, nouvelle espèce du Moyen Ogooué (Gabon)," *Bull. Mus. Natn. Hist. Natl. Paris* 4 (series 9), section B *Adansonia* 3: 229–37.

Hallwachs, W., 1986, "Agoutis: The Inheritors of Guapinol," in A. Estrada and T. H. Fleming, eds., *Frugivores and Seed Dispersal* (Dordrecht: Junk), pp. 286–304.

Hanski, Ikka, 1991, "North Temperate Dung Beetles," in Ikka Hanski and Yves Cambefort, eds., *Dung Beetle Ecology* (Princeton: Princeton University Press), pp. 255–78.

Hartman, Carl G., 1952, *Possums* (Austin: University of Texas Press).

Hartwig, Walter Carl, and Castor Cartelle, 1996, "A Complete Skeleton of the Giant South American Primate *Protopithecus*," *Nature* 381: 307–11.

Haynes, Gary, 1991, *Mammoths, Mastodonts, and Elephants* (New York: Cambridge University Press).

Heat-Moon, William Least, 1991, *PrairyErth* (Boston: Houghton Mifflin).

Heiser, Charles B., Jr., 1985, *Of Plants and People* (Norman: University of Oklahoma Press).

Herrera, Carlos M., 1985, "Determinants of Plant-Animal Coevolution: The Case of Mutualistic Dispersal of Seeds by Vertebrates," *Oikos* 44: 132–41.

______, 1986, "Vertebrate-Dispersed Plants: Why They Don't Behave the Way They Should," in A. Estrada and T. H. Fleming, eds., *Frugivores and Seed Dispersal* (Dordrecht: Junk), pp. 5–22.

______, 1989, "Frugivory and Seed Dispersal by Carnivorous Mammals, and Associated Fruit Characteristics, in Undisturbed Mediterranean Habitats," *Oikos* 55: 250–62.

______, 1989, "Vertebrate Frugivores and Their Interaction with Invertebrate Fruit Predators: Supporting Evidence from a Costa Rican Dry Forest," *Oikos* 54: 185–88.

Heymann, Edward W., and Gerald Hartmann, 1991, "Geophagy in Moustached Tamarins at the Rio Blanco, Peruvian Amazon," *Primates* 32: 533–37.

Hnatiuk, S. H., 1978, "Plant Dispersal by the Aldabran Giant Tortoise," *Oecologia* 36: 345–50.

Holdaway, R. N., 1996, "Arrival of Rats in New Zealand," *Nature* 384: 225–26.

Holdaway, R. N., and C. Jacomb, 2000, "Rapid Extinction of the Moas: Model, Test, and Implications," *Science* 287: 2250–54.

Holman, J. Alan, 1985, *Pleistocene Amphibians and Reptiles in North America* (New York: Oxford University Press).

Hori, T., et al., eds., 1997, *Ginkgo Biloba: A Global Treasure* (New York: Springer-Verlag).

Horn, Berthold K. P., 1978, "Dodo Apocrypha" (letter to the editor), *Science News* 113: 19.

Howe, Henry F., 1980, "Monkey Dispersal and Waste of a Neotropical Fruit," *Ecology* 6: 944–59.

______, 1984, "Implications of Seed Dispersal by Animals for Tropical Reserve Management," *Biological Conservation* 30: 261–81.

______, 1985, "Gomphothere Fruits: A Critique," *American Naturalist* 125: 853–65.

______, 1986, "Seed Dispersal by Fruit-Eating Birds and Mammals," in David R. Murray, ed., *Seed Dispersal* (Sydney: Academic Press), pp. 123–88.

______, 1989, "Scatter- and Clump-Dispersal and Seedling Demography," *Oecologia* 79: 417–26.

Howe, Henry F., and Gayle A. Vande Kerckhove, 1981, "Removal of Wild Nutmeg Crops by Birds," *Ecology* 62: 1093–1106.

Howe, Henry F., and Lynn C. Westley, 1988, *Ecological Relationships of Plants and Animals* (New York: Oxford University Press).

Hunter, J. Robert, 1989, "Seed Dispersal and Germination of *Enterolobium Cyclocarpum*," *Journal of Biogeography* 16: 369–78.

Izawa, Kosei, 1993, "Soil-Eating by *Alouatta* and *Ateles*," *International Journal of Primatology* 14: 229–41.

Jackson, Peter S. Wyse, Quentin C. B. Cronk, and John A. N. Parnell, 1988, "Notes on the Regeneration of Two Rare Mauritian Endemic Trees," *Tropical Ecology* 29: 98–106.

Jackson, Stephen T., and Chengyu Weng, 1999, "Late Quaternary Extinction of a Tree Species in Eastern North America," *Proceedings of the National Academy of Science* 96: 13847–52.

Janis, Christine, 1976, "The Evolutionary Strategy of the Equidae and the Origins of Rumen and Cecal Digestion," *Evolution* 30: 757–74.

Janson, Charles H., 1983, "Adaptation of Fruit Morphology to Dispersal Agents in a Neotropical Forest," *Science* 219: 187–89.

Janzen, Daniel H., 1970, "Herbivores and the Number of Tree Species in Tropical Forests," *American Naturalist* 104: 501–28.

______, 1971, "Escape of *Cassia Grandis* Beans from Predators in Time and Space," *Ecology* 52: 964–79.

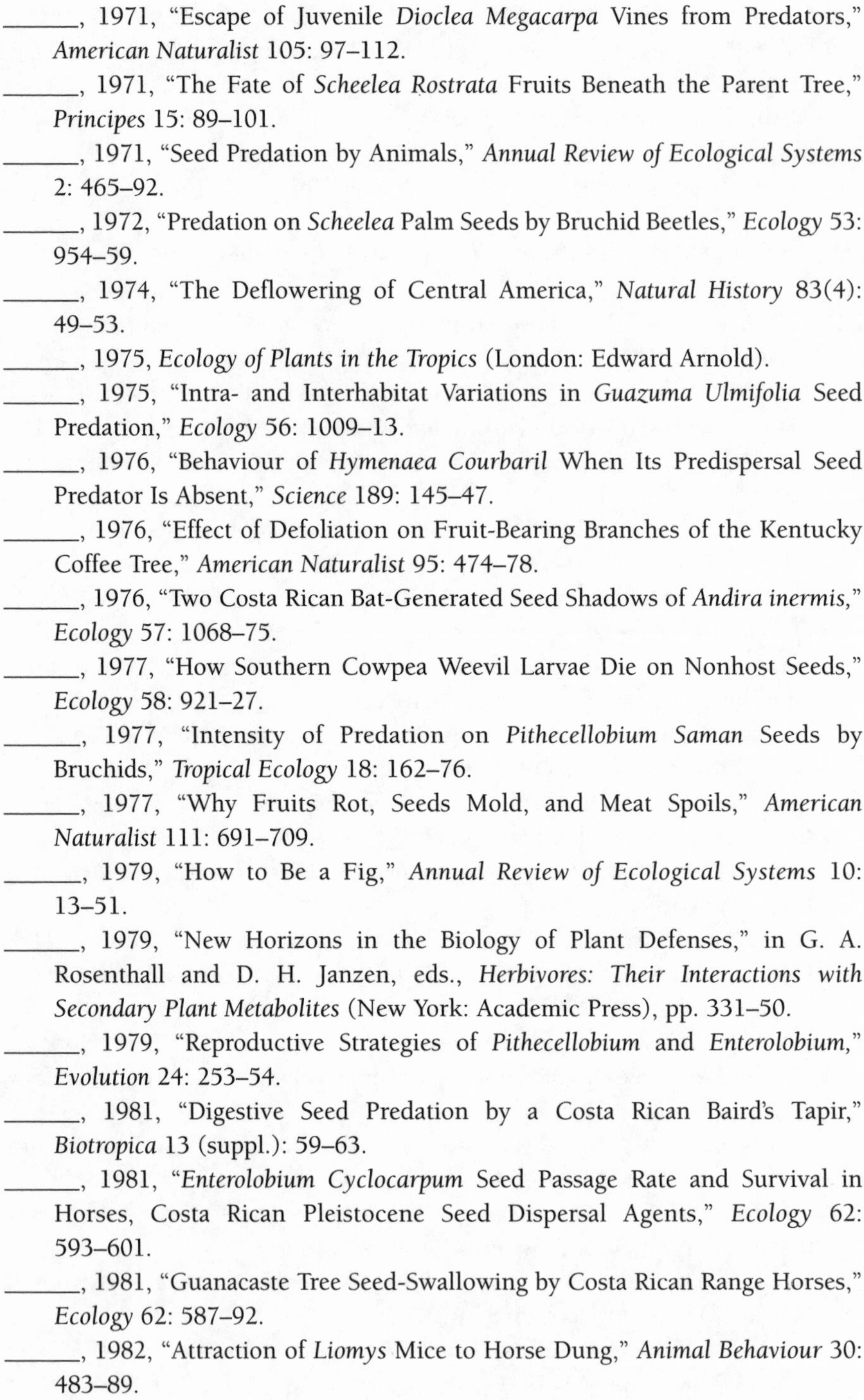

______, 1971, "Escape of Juvenile *Dioclea Megacarpa* Vines from Predators," *American Naturalist* 105: 97–112.

______, 1971, "The Fate of *Scheelea Rostrata* Fruits Beneath the Parent Tree," *Principes* 15: 89–101.

______, 1971, "Seed Predation by Animals," *Annual Review of Ecological Systems* 2: 465–92.

______, 1972, "Predation on *Scheelea* Palm Seeds by Bruchid Beetles," *Ecology* 53: 954–59.

______, 1974, "The Deflowering of Central America," *Natural History* 83(4): 49–53.

______, 1975, *Ecology of Plants in the Tropics* (London: Edward Arnold).

______, 1975, "Intra- and Interhabitat Variations in *Guazuma Ulmifolia* Seed Predation," *Ecology* 56: 1009–13.

______, 1976, "Behaviour of *Hymenaea Courbaril* When Its Predispersal Seed Predator Is Absent," *Science* 189: 145–47.

______, 1976, "Effect of Defoliation on Fruit-Bearing Branches of the Kentucky Coffee Tree," *American Naturalist* 95: 474–78.

______, 1976, "Two Costa Rican Bat-Generated Seed Shadows of *Andira inermis*," *Ecology* 57: 1068–75.

______, 1977, "How Southern Cowpea Weevil Larvae Die on Nonhost Seeds," *Ecology* 58: 921–27.

______, 1977, "Intensity of Predation on *Pithecellobium Saman* Seeds by Bruchids," *Tropical Ecology* 18: 162–76.

______, 1977, "Why Fruits Rot, Seeds Mold, and Meat Spoils," *American Naturalist* 111: 691–709.

______, 1979, "How to Be a Fig," *Annual Review of Ecological Systems* 10: 13–51.

______, 1979, "New Horizons in the Biology of Plant Defenses," in G. A. Rosenthall and D. H. Janzen, eds., *Herbivores: Their Interactions with Secondary Plant Metabolites* (New York: Academic Press), pp. 331–50.

______, 1979, "Reproductive Strategies of *Pithecellobium* and *Enterolobium*," *Evolution* 24: 253–54.

______, 1981, "Digestive Seed Predation by a Costa Rican Baird's Tapir," *Biotropica* 13 (suppl.): 59–63.

______, 1981, "*Enterolobium Cyclocarpum* Seed Passage Rate and Survival in Horses, Costa Rican Pleistocene Seed Dispersal Agents," *Ecology* 62: 593–601.

______, 1981, "Guanacaste Tree Seed-Swallowing by Costa Rican Range Horses," *Ecology* 62: 587–92.

______, 1982, "Attraction of *Liomys* Mice to Horse Dung," *Animal Behaviour* 30: 483–89.

______, 1982, "Cenizero Tree *(Pithecellobium Saman)* Delayed Fruit Development," *American Journal of Botany* 69: 1269–76.

______, 1982, "Differential Seed Survival and Passage Rates in Cows and Horses, Surrogate Pleistocene Dispersal Agents," *Oikos* 38: 150–56.

______, 1982, "Fruit Traits and Seed Consumption by Rodents of *Crescentia Alata*," *American Journal of Botany* 69: 1258–68.

______, 1982, "Fruits for Famished Mammoths," *Garden* 6(4): 12–14, 32.

______, 1982, "Horse Response to *Enterolobium Cyclocarpum* Fruit Crop Size," *Brenesia* 19/20: 209–19.

______, 1982, "How and Why Horses Open *Crescentia Alata* Fruits," *Biotropica* 14: 149–52.

______, 1982, "Natural History of Guacimo Fruits *(Guazuma Ulmifolia)* with Respect to Consumption by Large Mammals," *American Journal of Botany* 69: 1240–50.

______, 1982, "Removal of Seeds from Horse Dung by Tropical Rodents," *Ecology* 63: 1887–1900.

______, 1982, "Seeds in Tapir Dung," *Brenesia* 19/20: 129–35.

______, 1982, "Simulation of *Andira* Fruit Pulp Removal by Bats Reduces Seed Predation by Weevils," *Brenesia* 19/20: 165–70.

______, 1982, "Wild Plant Acceptability to a Captive Costa Rican Baird's Tapir," *Brenesia* 19/20: 99–128.

______, 1983, "Seasonal Change in Abundance of Large Nocturnal Dung Beetles in a Costa Rican Deciduous Forest and Adjacent Horse Pasture," *Oikos* 41: 274–83.

______, 1985, "How Fast and Why Do Germinating Guanacaste Seeds Die Inside Cows and Horses?" *Biotropica* 17(4): 322–25.

______, 1985, "On Ecological Fitting," *Oikos* 45: 308–10.

______, 1985, "*Spondias Mombin* Is Culturally Deprived in Megafauna-Free Forest," *Journal of Tropical Ecology* 1: 151–55.

______, 1986, "Chihuahuan Desert Nopaleras: Defaunated Big Mammal Vegetation," *Annual Review of Ecological Systems* 17: 595–636.

______, 1986, "Mice, Big Mammals, and Seeds: It Matters Who Defecates What Where," in A. Estrada and T. H. Fleming, eds., *Frugivores and Seed Dispersal* (Dordrecht: Junk), pp. 251–71.

______, 1988, "Complexity Is in the Eye of the Beholder," in F. Almeda and C. M. Pringle, eds., *Tropical Rainforests: Diversity and Conservation* (San Francisco: California Academy of Sciences), pp. 29–48.

Janzen, Daniel H., and Paul S. Martin, 1982, "Neotropical Anachronisms: The Fruits the Gomphotheres Ate," *Science* 215: 19–27.

Jefferson, Thomas, 1787/1955, *Notes on the State of Virginia* (Chapel Hill: University of North Carolina Press).

Johns, Timothy, 1990, *With Bitter Herbs They Shall Eat It* (Tucson: University of Arizona Press).

Jordano, Pedro, 1992, "Fruits and Frugivory," in M. Fenner, ed., *Seeds: The Ecology of Regeneration in Plant Communities* (Wallingford, England: Commonwealth Agricultural Bureau International), pp. 105–56.

______, 1995, "Angiosperm Fleshy Fruits and Seed Dispersers: A Comparative Analysis of Adaptation and Constraints in Plant-Animal Interactions," *American Naturalist* 145: 163–91.

Juretzek, W., 1997, "Recent Advances in *Ginkgo Biloba* Extract," in T. Hori et al., eds., *Ginkgo Biloba: A Global Treasure* (New York: Springer-Verlag), pp. 341–58.

Kammesheidt, Ludwig, 1999, "Forest Recovery by Root Suckers and Above-Ground Sprouts," *Journal of Tropical Ecology* 15: 143–57.

Kelly, D., and M. R. Ogle, 1990, "A Test of the Climate Hypothesis for Divaricate Plants," *New Zealand Journal of Ecology* 13: 51–61.

Klaus, G., et al., 1998, "Geophagy by Large Mammals at Natural Licks in the Rain Forest of the Dxanga National Park, Central African Republic," *Journal of Tropical Ecology* 14: 829–39.

Klimstra, W. D., and Frances Newsome, 1960, "Some Observations on the Food Coactions of the Common Box Turtle," *Ecology* 41: 639–47.

Kohn, J. R., and B. B. Casper, 1992, "Pollen-Mediated Gene Flow in *Cucurbita Foetidissima,*" *American Journal of Botany* 79: 57–62.

Konoshima, T., et al., 1995, "Anti-AIDS Agents from the Fruits of *Gleditsia Japonica* and *Gymnocladus Chinensis,*" *Journal of Natural Products–Lloydia* 58: 1372–77.

Krishnamani, R., and W. C. Mahaney, 2000, "Geophagy Among Primates: Adaptive Significance and Ecological Consequences," *Animal Behaviour* 59: 899–915.

Kruelen, F., 1985, "Lick Use by Large Herbivores," *Mammal Review* 15(3): 107–23.

Ladley, Jenny J., and Dave Kelly, 1996, "Dispersal, Germination, and Survival of New Zealand Mistletoes: Dependence on Birds," *New Zealand Journal of Ecology* 20(1): 69–79.

Lambert, Joanna E., and Paul A. Garber, 1998, "Evolutionary and Ecological Implications of Primate Seed Dispersal," *American Journal of Primatology* 45: 9–28.

Lamprey, H. F., G. Halevy, and S. Makacha, 1974, "Interactions Between *Acacia*, Bruchid Seed Beetles, and Large Herbivores," *East African Wildlife Journal* 12: 81–85.

Layne, Desmond R., 1996, "The Pawpaw," *HortScience* 31: 777–84.

Leopold, Aldo, 1949, *A Sand County Almanac* (New York: Oxford University Press).

Leopold, Donald J., et al., 1998, *Trees of the Central Hardwood Forests of North America* (Portland, Ore.: Timber Press).

Lieberman, Diana, and Milton Lieberman, 1987, "Notes on Seeds in Elephant Dung from Bia National Park, Ghana," *Biotropica* 19: 365–69.

MacPhee, Ross, 1997, "Digging Cuba: The Lesson of the Bones," *Natural History* 106(11): 50–54.

______, ed., 1999, *Extinctions in Near Time* (New York: Kluwer).

Mahaney, William C., 1999, "Chemistry, Mineralogy, and Microbiology of Termite Mound Soil Eaten by the Chimpanzees of the Mahale Mountains, Western Tanzania," *Journal of Tropical Ecology* 15: 565–88.

Mahaney, William C., et al., 1995, "Mountain Gorilla Geophagy," *International Journal of Primatology* 16: 475–87.

Mahaney, William C., et al., 1997, "Analysis of Geophagy Soils in Kibale Forest, Uganda," *Primates* 38: 159–76.

Mahaney, William C., et al., 2000, "Mineral and Chemical Analyses of Soils Eaten by Humans in Indonesia," *International Journal of Environmental Health Research* 10: 93–109.

Mandujano, Maria del Carmen, et al., 1996, "Reproductive Ecology and Inbreeding Depression in *Opuntia Rastrera* in the Chihuahuan Desert: Why Are Sexually Derived Recruitments So Rare?" *American Journal of Botany* 83: 63–70.

Margulis, Lynn, and Dorion Sagan, 1995, *What Is Life?* (New York: Simon & Schuster).

Martin, Paul S., 1966, "Africa and Pleistocene Overkill," *Nature* 212: 339–42.

______, 1967, "Overkill at Olduvai Gorge," *Nature* 215: 212–13.

______, 1969, "Wanted: A Suitable Herbivore," *Natural History* 78(2): 35–39.

______, 1970, "Pleistocene Niches for Alien Animals," *BioScience* 20: 218–21.

______, 1975, "Vanishings, and Future, of the Prairie," *Geoscience and Man* 10: 39–47.

______, 1990, "Forty Thousand Years of Extinctions on the 'Planet of Doom,'" *Palaeogeography, Palaeoclimatology, Palaeoecology* 82: 182–201.

______, 1990, "Thinking Like a Canyon: Wild Ideas and Wild Burros," in Robert H. Webb, ed., *A Century of Environmental Change in Grand Canyon* (Tucson: University of Arizona Press), pp. 82–83.

______, 1992, "The Last Entire Earth," *Wild Earth*, winter, pp. 29–32.

Martin, Paul S., and David Burney, 1999, "Bring Back the Elephants!" *Wild Earth* 9(1): 57–64.

Martin, Paul S., and David W. Steadman, 1999, "Prehistoric Extinctions on Islands and Continents," in Ross MacPhee, ed., *Extinctions in Near Time* (New York: Kluwer), pp. 17–55.

Martin, Paul S., and Christine R. Szuter, 1999, "War Zones and Game Sinks in Lewis and Clark's West," *Conservation Biology* 13: 36–45.

Mason, C. T., 1975, [title unavailable], *Madroño* 23(3): 105–8.

Matheny, William A., 1931, *Seed Dispersal* (Ithaca, N.Y.: Slingerland Comstock).

McDougall, W. B., and Omer E. Sperry, 1951, *Plants of Big Bend National Park* (Washington, D.C.: USGPO).

McGlone, M. S., and B. D. Clarkson, 1993, "Ghost Stories: Moa, Plant Defences, and Evolution in New Zealand," *Tuatara* 32: 1–18.

McGlone, M. S., and C. J. Webb, 1981, "Selective Forces Influencing the Evolution of Divaricating Plants," *New Zealand Journal of Ecology* 4: 20–28.

Michener, David C., 1986, "Phenotypic Instability in *Gleditsia Triacanthos*," *Brittonia* 38: 360–61.

Miller, Maxine F., 1993, "Is It Advantageous for Acacia Seeds to Be Eaten by Ungulates?" *Oikos* 66: 364–68.

______, 1996, "Dispersal of *Acacia* Seeds by Ungulates and Ostriches in an African Savanna," *Journal of Tropical Ecology* 12: 345–46.

Miller, Scott E., et al., 1981, "Reevaluation of Pleistocene Scarab Beetles from Rancho La Brea, California," *Proceedings of the Entomological Society of Washington* 83: 625–30.

Milton, S. J., et al., 1990, "The Distribution of Epizoochoric Plant Species," *Journal of Biogeography* 17: 25–34.

Mueller, R. F., 1990, "Floral Legacies of the Megafauna," *Earth First*, February, p. 23.

Murray, David R., ed., 1986, *Seed Dispersal* (Sydney: Academic Press).

Myers, Judith H., and Dawn Bazely, 1991, "Thorns, Spines, Prickles, and Hairs," in D. W. Tallamy and M. J. Raupp, eds., *Phytochemical Induction by Herbivores* (New York: Wiley), pp. 325–44.

Nabhan, Gary Paul, 1987, *Gathering the Desert* (Tucson: University of Arizona Press).

______, 1987, "Origins of Neotropical Horticulture Following Megafaunal Extinctions: Did Mesoamericans Disperse and Select Anachronistic Fruits?" presentation at the ethnobiology conference, March, Gainesville, Florida.

______, 1989, *Enduring Seeds* (San Francisco: North Point Press).

Newson, Lee A., S. David Webb, and James S. Dunbar, 1993, "History and Geographic Distribution of *Cucurbita Pepo* Gourds in Florida," *Journal of Ethnobiology* 12: 75–97.

Niklas, Karl J., 1997, *The Evolutionary Biology of Plants* (Chicago: University of Chicago Press).

Noble, J. C., 1975, "The Effects of Emus on the Distribution of the Nitre Bush," *Journal of Ecology* 63: 979–84.

Norman, Eliane M., and David Clayton, 1986, "Reproductive Biology of Two Florida Pawpaws," *Bulletin of the Torrey Botanical Club* 113: 16–22.

Norman, Eliane M., Kathleen Rice, and Steven Cochran, 1992, "Reproductive Biology of *Asimina Parviflora*," *Bulletin of the Torrey Botanical Club* 119: 1–5.

Nwokolo, E., 1996, "African Breadfruit," in E. Nwokolo and J. Smartt, eds., *Food and Feed from Legumes and Oilseeds* (London: Chapman & Hall), pp. 345–54.

______, 1996, "Bottle Gourd, Buffalo Gourd, and Other Gourds," in E. Nwokolo and J. Smartt, eds., *Food and Feed from Legumes and Oilseeds* (London: Chapman & Hall), pp. 290–97.

Nwokolo, E., and J. Smartt, eds., 1996, *Food and Feed from Legumes and Oilseeds* (London: Chapman & Hall).

O'Dowd, Dennis J., and A. Malcolm Gill, 1986, "Seed Dispersal Syndromes in Australian Acacia," in David R. Murray, ed., *Seed Dispersal* (Sydney: Academic Press), pp. 87–121.

Oliveira, Paulo S., et al., 1995, "Seed Cleaning by Ants Facilitates Germination in *Hymenaea Courbaril*," *Biotropica* 27: 518–22.

Osman, M. A., and C. W. Weber, 1995, "Isolation and Properties of Proteinase-Inhibitor from Buffalo Gourd Seed," *FASEB Journal* 9: A748.

Owadally, A. W., 1979, "The Dodo and the Tambalacoque Tree" (letter to the editor), *Science* 203: 1363–64.

Owen-Smith, R. Norman, 1987, "Pleistocene Extinctions: The Pivotal Role of Megaherbivores," *Paleobiology* 13: 351–62.

______, 1988, *Megaherbivores* (Cambridge: Cambridge University Press).

Patterson, Bryan, 1975, "The Fossil Aardvarks," *Bulletin of the Museum of Comparative Zoology* 147(5): 185–237.

Payne, Junaidi, 1990, "Rarity and Extinctions of Large Mammals in Malaysian Rainforests," in Son-Kheong Yap, Su Win Lee, and the Malayan Nature Society, eds., *In Harmony with Nature: Proceedings of the International Conference on Tropical Biodiversity* (Kuala Lumpur: Malayan Nature Society), pp. 310–11.

Peattie, Donald Culross, 1950, *A Natural History of Western Trees* (Boston: Houghton Mifflin).

Platt, Rutherford, 1952, *American Trees* (New York: Dodd, Mead).

Pratt, Thane K., 1982, "Pleistocene Seed Dispersal," *Science* 216: 6.

Profet, M., 1992, "Pregnancy Sickness as Adaptation: A Deterrent to Maternal Ingestion of Teratogens," in J. H. Barkow, *The Adapted Mind* (New York: Oxford University Press), pp. 327–65.

Quammen, David, 1996, *The Song of the Dodo* (New York: Scribner).

______, 1998, "Planet of Weeds," *Harper's*, October, pp. 57–69.

Redford, Kent H., 1992, "The Empty Forest," *BioScience* 42: 412–22.

Rehr, S. S., E. Arthur Bell, Daniel H. Janzen, and Paul P. Feeny, 1973, "Insecticidal Amino Acids in Legume Seeds," *Biochemical Systematics* 1: 63–67.

Reynolds, Harold C., 1945, "Some Aspects of the Life History and Ecology of the Opossum in Central Missouri," *Journal of Mammalogy* 26: 361–78.

Rick, Charles M., and Robert I. Bowman, 1960, "Galápagos Tomatoes and Tortoises," *Evolution* 15: 407–17.

Rogers, Julia Ellen, 1917, *Trees Worth Knowing* (Garden City, N.Y.: Doubleday).

Rogers, Lynn L., and Roger D. Applegate, 1983, "Dispersal of Fruit Seed by Black Bears," *Journal of Mammalogy* 64(2): 310–11.

Rohner, Christoph, and David Ward, 1999, "Large Mammalian Herbivores and the Conservation of Arid *Acacia* Stands in the Middle East," *Conservation Biology* 13: 1162–71.

Rudwick, Martin J. S., 1997, *Georges Cuvier, Fossil Bones, and Geological Catastrophes* (Chicago: University of Chicago Press).

Rust, R. W., and R. R. Roth, 1981, "Seed Production and Seedling Establishment in the Mayapple," *American Midland Naturalist* 105: 51–60.

Sastre, J., et al., 1998, "A *Ginkgo Biloba* Extract Prevents Mitochondrial Aging by Protecting Against Oxidative Stress," *Free Radical Biology and Medicine* 24(2): 298–304.

Schambach, Frank F., 2000, "Spiroan Traders, the Sanders Site, and the Plains Interaction Sphere," *Plains Anthropologist* 45(171): 7–33.

Schnabel, Andrew, and J. L. Hamrick, 1990, "Organization of Genetic Diversity Within and Among Populations of *Gleditsia Triacanthos*," *American Journal of Botany* 77: 1060–69.

Schnabel, Andrew, Roger H. Laushman, and J. L. Hamrick, 1991, "Comparative Genetic Structure of Two Cooccurring Tree Species, *Maclura Pomifera* and *Gleditsia Triacanthos*," *Heredity* 67: 357–64.

Schnabel, Andrew, J. D. Nason, and J. L. Hamrick, 1998, "Understanding the Population Genetic Structure of *Gleditsia Triacanthos*," *Molecular Ecology* 7: 819–32.

Schnabel, Andrew, and Jonathan F. Wendel, 1998, "Cladistic Biogeography of *Gleditsia* (Leguminosae) Based on *NDHF* and *RPL16* Chloroplast Gene Sequences," *American Journal of Botany* 85: 1753–65.

Seligmann, Jean, and Bill Hart, 1988, "Where the Camels Roam," *Newsweek*, 7 November, p. 23.

Simpson, B. B., 1972, *Mesquite* (Washington, D.C.: Institute of Ecology).

Sinclair, A. R. E., and Peter Arcese, eds., 1995, *Serengeti II* (Chicago: University of Chicago Press).

Smith, Jeffrey L., and Janice V. Perino, 1981, "Osage Orange: History and Economic Uses," *Economic Botany* 35: 24–41.

Smythe, N., 1986, "Competition and Resource Partitioning in the Guild of Neotropical Terrestrial Frugivorous Mammals," *Annual Review of Ecology and Systematics* 17: 169–88.

Sork, Victoria L., 1987, "Effects of Predation and Light on Seedling Establishment in *Gustavia Superba*," *Ecology* 68: 1341–50.

Soulé, Michael, and Reed Noss, 1998, "Rewilding and Biodiversity: Complementary Goals for Continental Conservation," *Wild Earth* 8(3): 18–28.

Steadman, David W., and Paul S. Martin, 1984, "Extinction of Birds in the Late Pleistocene of North America," in Paul S. Martin and Richard G. Klein, eds., *Quaternary Extinctions* (Tucson: University of Arizona Press), pp. 466–77.

Stenseth, Nils C., 1983, "Grasses, Grazers, Mutualism and Coevolution: A Comment About Handwaving in Ecology," *Oikos* 41(1): 152–59.

Stephens, H. A., 1980, *Poisonous Plants of the Central United States* (Lawrence: University Press of Kansas).

Stevens, C. Edward, and Ian D. Hume, 1995, *Comparative Physiology of the Vertebrate Digestive System* (Cambridge: Cambridge University Press).

Stocker, G. C., and A. K. Irvine, 1983, "Seed Dispersal by Cassowaries in North Queensland's Rainforests," *Biotropica* 15(3): 170–76.

Stone, Richard, 1998, "A Bold Plan to Recreate a Long-Lost Siberian Ecosystem," *Science* 282: 31–33.

______, 1999, "Cloning the Woolly Mammoth," *Discover,* April, pp. 56–63.

Sun, Marjorie, 1988, "Costa Rica's Campaign for Conservation," *Science* 239: 1366–69.

Swain, Tony, 1976, "Angiosperm-Reptile Coevolution," in A. D. Bellairs and C. B. Cox, eds., *Morphology and Biology of Reptiles,* pp. 107–22.

Taulman, James F., and James H. Williamson, 1994, "Food Preferences of Captive Wild Raccoons," *Canadian Field-Naturalist* 108: 170–75.

Tchamba, Martin N., and Prosper M. Seme, 1993, "Diet and Feeding Behavior of the Forest Elephant in the Santchou Reserve, Cameroon," *African Journal of Ecology* 31: 165–71.

Temple, Stanley A., 1977, "Plant-Animal Mutualism: Coevolution with Dodo Leads to Near Extinction of Plant," *Science* 197: 885–86.

Tozer, Eliot, 1980, "The Honey Locust," *Horticulture,* September, pp. 48–52.

Traverse, Alfred, 1988, "Plant Evolution Dances to a Different Beat," *Historical Biology* 1: 277–301.

Traveset, Anna, 1990, "Post-Dispersal Predation of *Acacia Farnesiana* Seeds by *Stator Vachelliae* (Bruchidae) in Central America," *Oecologia* 84: 506–12.

Turner, Nancy J., and Adam F. Sczawinski, 1991, *Common Poisonous Plants and Mushrooms of North America* (Portland, Ore.: Timber Press).

Tutin, Caroline E. G., et al., 1991, "A Case Study of a Plant-Animal Relationship: *Cola Lizae* and Lowland Gorillas in the Lopé Reserve, Gabon," *Journal of Tropical Ecology* 7: 181–99.

Tutin, Caroline E. G., et al., 1996, "Protecting Seeds from Primates: Examples from *Diospyros* Spp. in the Lopé Reserve, Gabon," *Journal of Tropical Ecology* 12: 371–84.

Van der Pijl, L., 1972, *Principles of Dispersal in Higher Plants* (New York: Springer-Verlag).

Van Froenendael, J. M., et al., 1996, "Comparative Ecology of Clonal Plants," *Philosophical Transactions of the Royal Society of London*, series B, 351: 1331–39.

Vasek, Frank C., 1980, "Creosote Bush: Long-Lived Clones in the Mojave Desert," *American Journal of Botany* 67: 246–55.

Wada, Keiji, and Masanobu Haga, 1997, "Food Poisoning by *Ginkgo Biloba* Seeds," in T. Hori et al., eds., *Ginkgo Biloba: A Global Treasure* (New York: Springer-Verlag), pp. 309–19.

Werthner, William B., 1935, *Some American Trees* (New York: Macmillan).

Western, David, 1997, *In the Dust of Kilimanjaro* (Washington, D.C.: Island Press).

Wheelwright, Nathaniel T., and Gordon H. Orians, 1982, "Seed Dispersal by Animals: Contrasts with Pollen Dispersal, Problems of Terminology, and Constraints of Coevolution," *American Naturalist* 119: 403–13.

Whitaker, A. H., 1987, "The Roles of Lizards in New Zealand Plant Reproductive Strategies," *New Zealand Journal of Botany* 25: 315–28.

White, Lee J. T., Caroline E. G. Tutin, and Michael Fernandez, 1993, "Group Composition and Diet of Forest Elephants in the Lopé Reserve, Gabon," *African Journal of Ecology* 31: 181–99.

White, Peter S., 1988, "Prickle Distribution in *Aralia Spinosa*," *American Journal of Botany* 75(2): 282–85.

Whitney, Kenneth D., et al., 1998, "Seed Dispersal by *Ceratogymna* Hornbills in the Dja Reserve, Cameroon," *Journal of Tropical Ecology* 14: 351–71.

Wiley, Andrea S., and Solomon H. Katz, 1998, "Geophagy in Pregnancy: A Test of a Hypothesis," *Current Anthropology* 39: 532–45.

Williams, Nigel, 1998, "Study Finds 10% of Tree Species Under Threat," *Science* 281: 1426.

Willson, Mary F., 1993, "Mammals as Seed-Dispersal Mutualists in North America," *Oikos* 67: 159–76.

______, 1997, "Effects of Birds and Bears on Seed Germination," *Oikos* 80: 89–95.

Willson, Mary F., A. K. Irvine, and Neville G. Walsh, 1989, "Vertebrate Dispersal Syndromes in Some Australian and New Zealand Communities," *Biotropica* 21(2): 133–47.

Willson, Mary F., and Douglas W. Schemske, 1980, "Pollinator Limitation, Fruit Production, and Floral Display in Pawpaw," *Bulletin of the Torrey Botanical Club* 107: 401–8.

Wilson, Don E., and Daniel H. Janzen, 1972, "Predation on *Scheelea* Palm Seeds by Bruchid Beetles," *Ecology* 53: 954–59.

Wilson, R. T., 1989, *Ecophysiology of the Camelidae and Desert Ruminants* (New York: Springer-Verlag).

Witmer, Mark C., 1991, "The Dodo and the Tambalacoque Tree: An Obligate Mutualism Reconsidered," *Oikos* 61: 133–37.

Worth, C. Brooke, 1975, "Virginia Opossums as Disseminators of the Common Persimmon," *Journal of Mammology* 56: 517.

Yamashita, Carlos, 1997, "Anodorhynchus Macaws as Followers of Extinct Megafauna: An Hypothesis," *Ararajuba* 5(2): 176–82.

Yumoto, Takakazu, and Tamaki Maruhashi, 1995, "Seed Dispersal by Elephants in a Tropical Rain Forest in Kahuzi-Biega National Park, Zaire," *Biotropica* 27(4): 526–30.

Zimov, Sergei A., et al., 1995, "Steppe-Tundra Transition: A Herbivore-Driven Biome Shift at the End of the Pleistocene," *American Naturalist* 146: 765–94.

INDEX